Green Energy and Technology

Climate change, environmental impact and the limited natural resources urge scientific research and novel technical solutions. The monograph series Green Energy and Technology serves as a publishing platform for scientific and technological approaches to “green”—i.e. environmentally friendly and sustainable—technologies. While a focus lies on energy and power supply, it also covers “green” solutions in industrial engineering and engineering design. Green Energy and Technology addresses researchers, advanced students, technical consultants as well as decision makers in industries and politics. Hence, the level of presentation spans from instructional to highly technical.

Indexed in Scopus.

Indexed in Ei Compendex.

Namrata Manohar · Mousmi Ajay Chaurasia ·
Stefan Mozar · Chia-Feng Juang

Machine Learning Applications in Renewable Energy

Springer

Namrata Manohar
HCL
Hyderabad, Telangana, India

Mousmi Ajay Chaurasia
Koneru Lakshmaiah Education Foundation
Hyderabad, Telangana, India

Stefan Mozar
Dynexsys Pty. Ltd.
Glenwood, Australia

Chia-Feng Juang
Department of Electrical Engineering
National Chung Hsing University
Taichung, Taiwan

ISSN 1865-3529 ISSN 1865-3537 (electronic)
Green Energy and Technology
ISBN 978-981-97-9938-1 ISBN 978-981-97-9939-8 (eBook)
https://doi.org/10.1007/978-981-97-9939-8

This Springer imprint is published by the registered company Springer Nature Singapore Pte Ltd.
The registered company address is: 152 Beach Road, #21-01/04 Gateway East, Singapore 189721, Singapore

Contents

Chapter 1
Renewable Energy Technologies (RET)—Society and Government Perspective

The Chapter presents the need for Renewable Energy Technologies (RET) in the context of providing a solution for depletion of conventional sources, protection of environment and enhancing the economical situation of a country by way of providing employment opportunities for many people may be as employees in various roles or initiating their own enterprise. The Chapter presents statistics on the change in energy consumption over the past few decades from conventional to renewable energies, the future scenario in view of advancements in technologies and also about the employment status past, present and future. The importance of standards and how modern technologies are enhancing the successful implementation of RET is presented. The role of the public and government and the various financial schemes provided by governments is highlighted. A few modern applications and those under development would enhance the standard of living. The statistics and situation of the various aspects before and after COVID-19 pandemic, and future effects are discussed, for overall benefit of one and all.

1.1 Introduction

Electrical Energy is one of the most vital energy sources for conducting our daily activities There was a time when only Industries and Offices could not operate for even an hour without this energy. But, today the day has come where each individual human will go berserk if the energy is not available for even a second, due to the use of several modern gadgets as mobiles, laptops etc. Thus, there is a huge increase in the power consumption from Kilo-Watts in Industrial plants to a few milli-Watts by each individual. The areas of usage have also increased, where about a decade and a half back it was only industrial, residential and commercial; now it is industrial, residential, commercial, agricultural, education, banking, restaurants, transport,

N. Manohar et al., *Machine Learning Applications in Renewable Energy*, Green Energy and Technology, https://doi.org/10.1007/978-981-97-9939-8_1

entertainment, personal etc. The areas have increased in number, and quantity of electricity consumption from Organization level to individual level. As each person is using minimum two gadgets—mobile phone and a computer system or Laptop.

Figure 1.1 shows the World Population Distribution in Rural and Urban Areas Up to the Mid of the 21th Century [1]. It can be seen that there is a large migration of population from rural to urban. We know that power consumption is more in Urban areas than rural areas, and this may continue to be the scenario at least for another century. Hence from Fig. 1.1 we can say that as the population in moving towards urban areas, therefore the electricity consumption in urban areas will also be in the rise.

It may also be noted that even in rural areas the electricity power consumption at residential and individual level would be in the rise due to proliferation of electronic gadgets. Figure 1.2: Annual Energy Usage by type of purpose in Rural and Urban Areas Fig. 1.3: Annual Energy Consumption by type of fuel in Rural and Urban Areas

The advancements in various new technologies—including the full diffusion of information, sensing and communication technologies, automation, more productive resource technologies, and health technologies—will have the potential to reshape

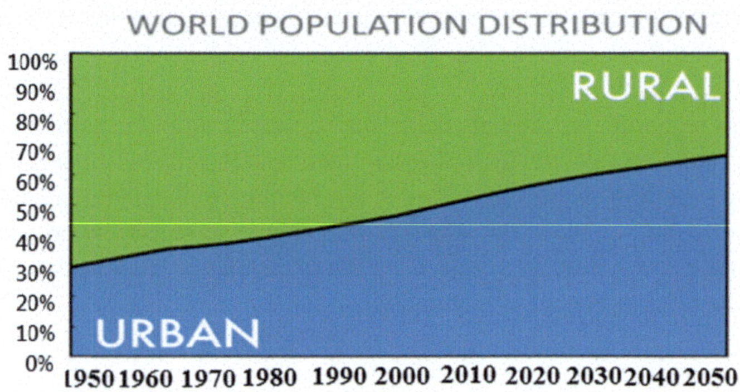

Fig. 1.1 World population distribution—rural and urban areas upto the mid of the 21th century

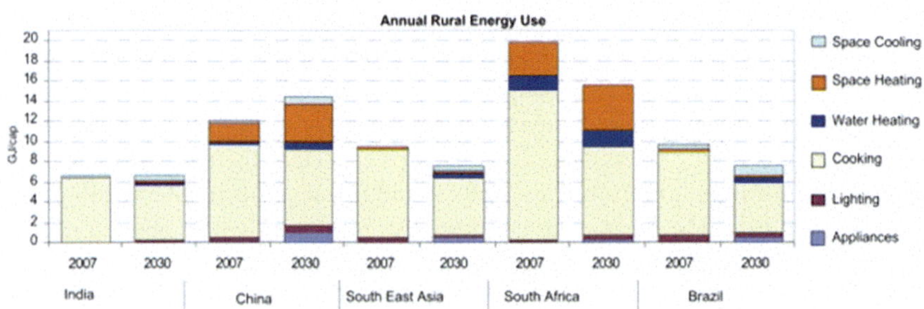

Fig. 1.2 Annual energy usage by the type of purpose in urban and rural areas

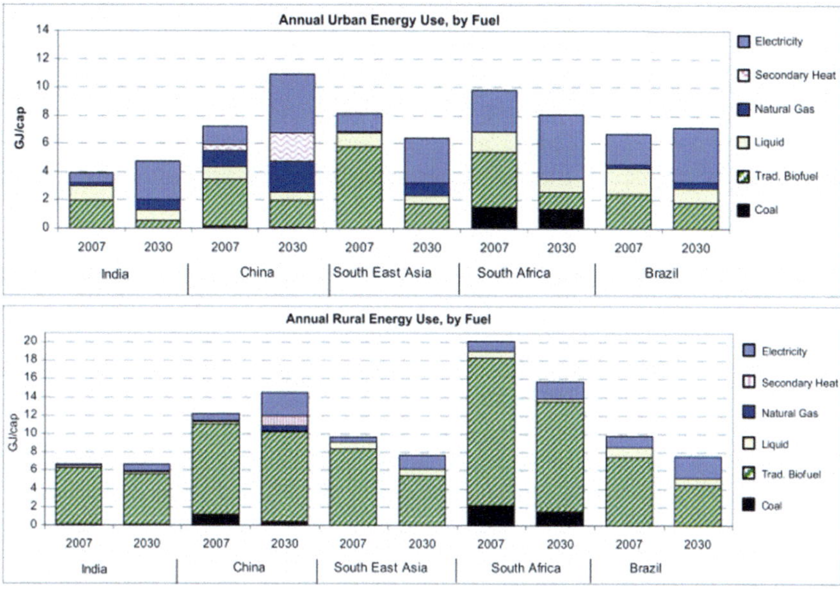

Fig. 1.3 Annual energy consumption by type of fuel in rural and urban areas

economic and social options. The pace at which these new technologies develop continues to accelerate at an exponential rate whilst technology costs tumble and the technologies become ubiquitous. Furthermore, the combinative effect of these technologies is creating the environment for fundamental change. For example, fully un-manned Electric Vehicles (EVs) are fast becoming a reality. The combinative effect of battery technology, GPS, machine learning and analytics have created the right environment to accelerate what was previously seen as niche development or a longer-term change. As these technologies mature, combine and are deployed across the economy, in the broadest sense we will see major changes in the energy industry, due to evolvement of:

- Smart cities,
- Automation
- Increase in digital productivity
- Energy efficiency
- Demand for zero carbon emissions
- Wind and solar, and integrated grid/storage
- Electric vehicles (EV)

In short, the dominant technologies that will drive the energy industry in the Grand Transition will be very different from those employed during the past 45 years.

All Governments around the world have introduced policies and regulations to support the development and deployment of renewable energy technology. Incentives and subsidies play a crucial role in improving renewable energy competitiveness.

Figure 1.4 shows the statistics of Number of Renewable Energy Policies (a) Type Wise (b) Regional and (c) End User Segment [2]. They provide financial support over an extended period. Public investments and policy processes are anticipated to grow at a high Compound Annual Growth Rate (CAGR) during the forthcoming period. Education and outreach initiatives, along with research, development, and deployment (RD&D) efforts, are expected to experience notable growth by 2029. This highlights the increasing focus on knowledge dissemination, awareness-building, and innovation in the renewable energy market. Renewable Energy Technologies (RET) are the need of the hour. They have a key role in the present advanced technology society for several reasons but the two main reasons are—depletion of natural resources for generation of electricity and second vast availability of Renewable Energy Sources (RES). RET make energy available at the point of usage making it easy for installation, operation and maintenance and also reduction in transmission losses. RET have made energy generation compact and economical. This feature is very compatible with the technological advances in various application areas and equipment from major to minor. Considering the major aspect we have the electricity grid which is becoming nuclear in size and also hybrid which is possible for efficient operation due to the advancements in Power Electronics Devices, Process Control Instrumentation and Information Technology advances as Artificial Intelligence, IOT.

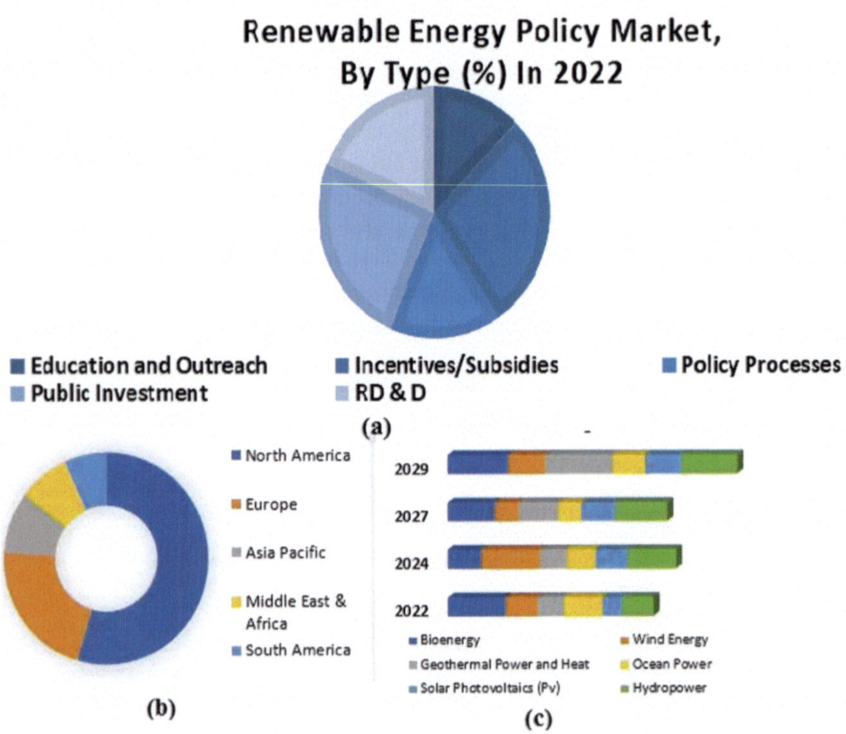

Fig. 1.4 Number of renewable energy policies **a** Type wise **b** Regional and **c** End user segment

On the major application front we have electric trains and electric vehicles. Minor aspect is that all electrical devices and equipment are becoming portable and operating on chargeable batteries requiring the need to have electric generating stations which are small, reliable and economical.

Technological requirement is all said and done but no project can be successful without effective management. In implementing RET both public and Government have a role to play. This service area provides a platform for the public and Government to mutually exchange and share the energy load requirements considering the technical and financial benefits of all stakeholders. Figure 1.5 shows the scenario of Grid and RES integrated to supply power to a load, as on need and availability basis.

Let P_s—Power Supply by the Conventional Sources in KW.

P_d—Power Demand by the load in KW

P_{res}—Power generated by RES in KW

The conditions that exist are

i. If Pres > Pd the excess power (Pres− Pd) is put on to the grid (1.1)

ii. If Pres < Pd the shortfall power (Pd − Pres) is drawn from the grid (1.2)

Implementation of RET has now become more easy, with the developments in all branches of Engineering—Electrical, Electronics, Computers, Civil, Mechanical and Chemical. Operation has become effective with the advent of Microcontrollers, SCADA, various digital measuring devices and Control Devices and Algorithms.

1.1.1 Renewable Energy–Sources and Application Areas

Before we delve into the detail aspects of Renewable Energy Technology, let us fist recap the various renewable energy sources and their application areas, so that the various aspects presented can be understood in the right context. Figure 1.6 [3] indicates the various renewable energy sources and Table 1.1 and Fig. 1.7 the application areas and systems. The application areas and systems being extensive the important and most common are presented.

1.2 Renewable Energy Technology—Need of the Hour

Technology advancements, increase in literate community, awareness of developments in society has resulted in increased standard of living and economy, thereby energy needs are increasing, but the current energy generation is not sufficient to

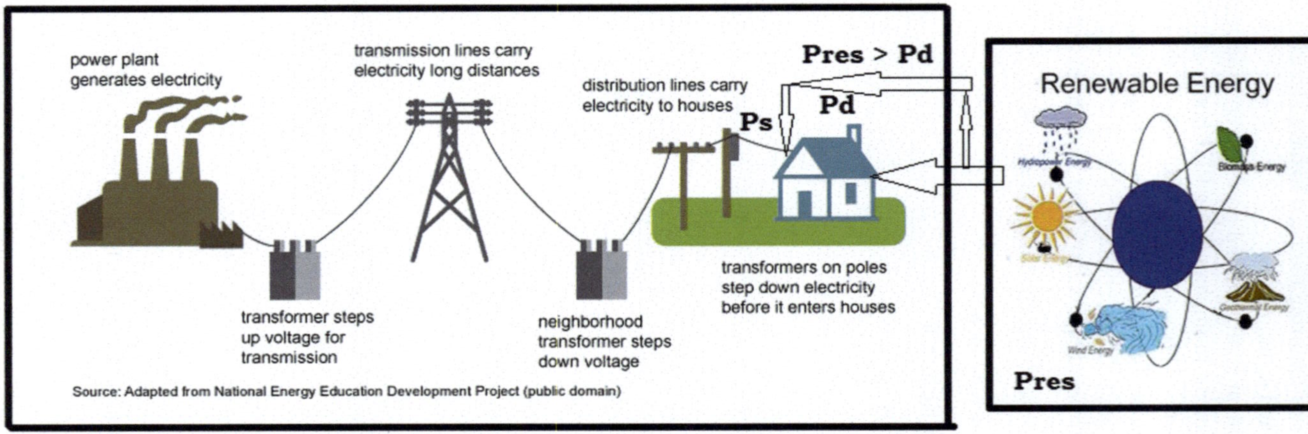

Fig. 1.5 Integration and exchange of power between grid and renewable energy sources

Fig. 1.6 Renewable energy sources

Table 1.1 Renewable energy application areas and systems

Application area	Application systems
Residential	Water pumping
Industrial	Water heaters
Transport	Solar cookers
Educational institutions	Lightening
Commercial	Room heaters
Hospitals	Industrial motors
Agricultural	Domestic appliances
Hotels	Vehicles

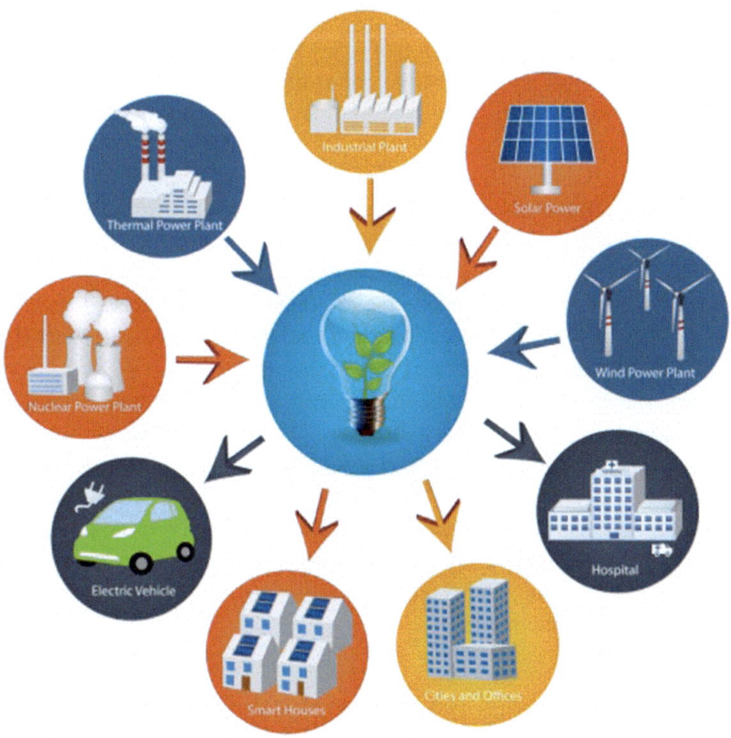

Fig. 1.7 Renewable energy application areas

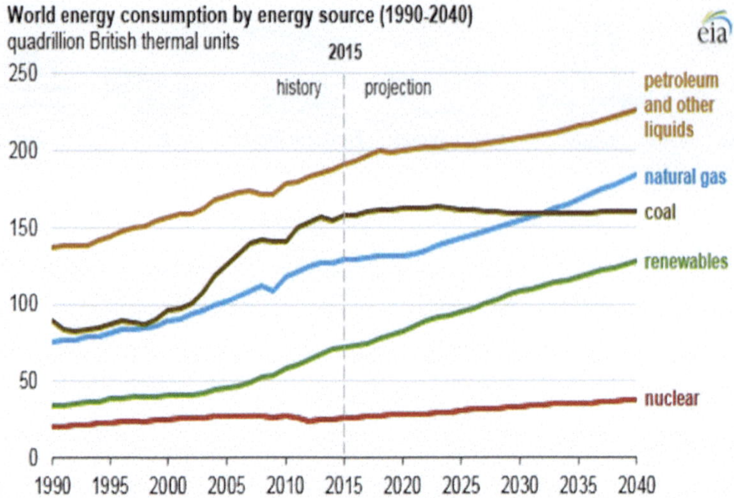

Fig. 1.8 Increase in consumption of electrical energy over the next two decades

meet our requirements.. The geographical extent of energy usage in terms of Kilometers is also increasing with expansion of cities, towns and villages. But as is known setting up Electric Power System is time consuming and cost intensive. Renewable Energy [4] therefore would give the solution for energy at door-step and power for all anywhere and everywhere.

RET enable us to create various forms of energy as electricity, heat and fuel from renewable sources. Renewable energy has huge potential to provide solution for energy crisis and is the key factor to food and energy security.

Figure 1.8 shows the increase in consumption of electrical energy over the next two decades [5–7]. The projections are based on the actual consumption during the past two decades. It can be observed that the rise in renewable is steeper than the other forms of energy.

The renewable energy sources as wind, ocean, tidal can be undertaken only by Government and big Organizations as they are expensive and complex in terms of technical aspects and financial. The renewable energy sources which can be installed by individuals and Groups as Housing Colonies, Commercial establishments, industries, Educational Institutions etc. is the solar system which is easy and cost effective. The main advantage of RET is almost zero operational cost.

The factors necessitating the need for renewable energy [8] and the situations for them arising are presented in Table 1.2.

Table 1.2 Factor→scenario of renewable energy

Factor	Scenario
1. Development of equipment's and processes to harness energy from renewable energy sources. Need for energy generation sources at the point of usage. Possibility of power quality control by way of small and compact power electronic devices	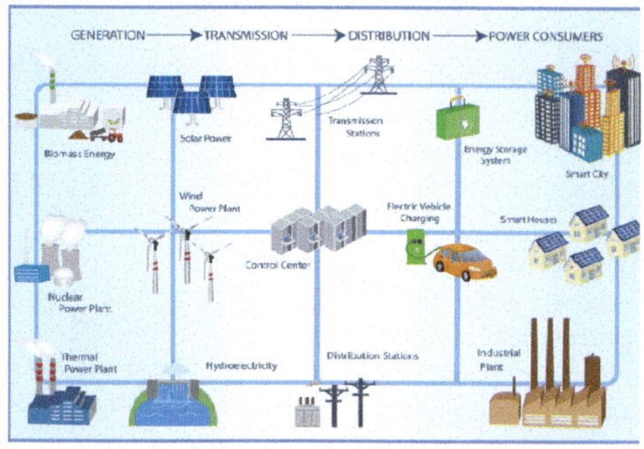
2. Depletion of non-renewable resources	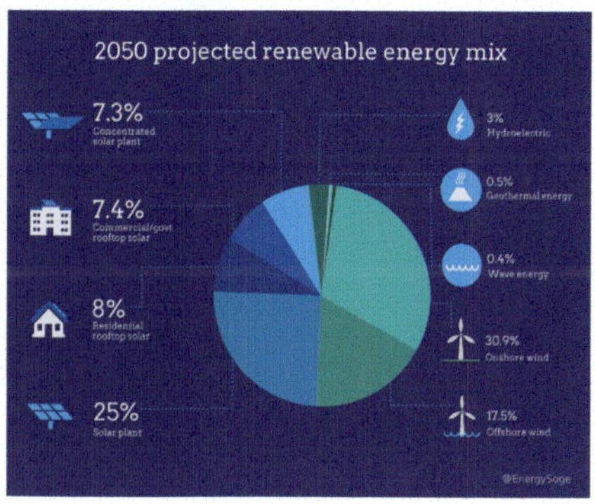
3. De-centralisation of electricity Sector—The change in the management of electric sector from centralised to decentralised, is providing more opportunities and avenues for individuals to install their own on-site power plants	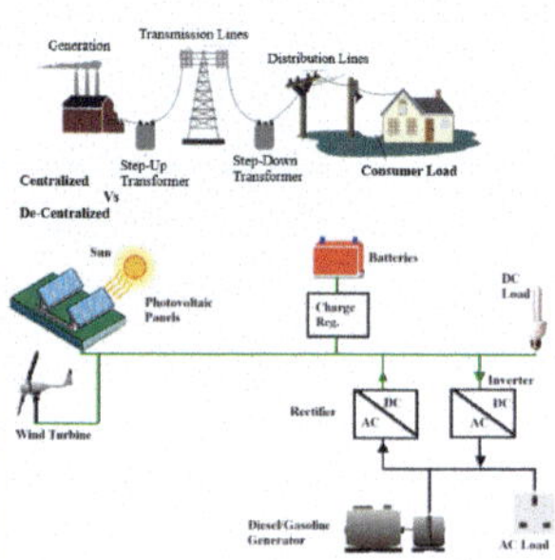

(continued)

Table 1.2 (continued)

Factor	Scenario
4. Development of various domestic gadgets as Ovens, washing machines, electric stoves, small kitchen geysers. personal gadgets as cell phones, laptops, e-pads, electric vehicles. Office, commercial, industrial and various service sector equipments as Computers, communication equipment, lifts, elevators, water purifies, water pumping, laboratories, research equipment etc.	  
5. Increase in air pollution -	
6. Need for cost-reduction, Increasing utility savings	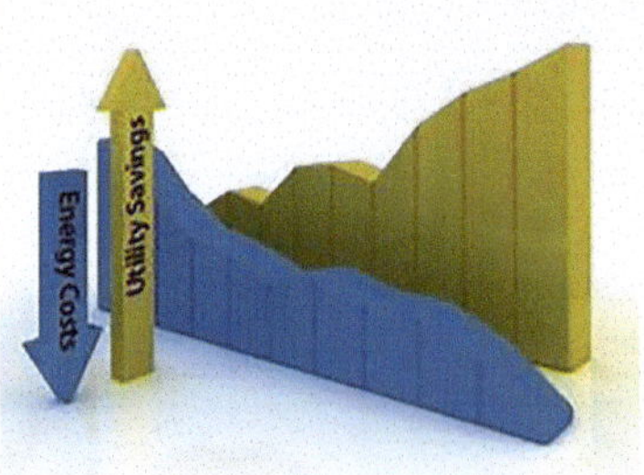

(continued)

Table 1.2 (continued)

Factor	Scenario
7. Increased competitive market	
8. Extention of inhabited areas as cities, villages over several stretches	
9. Government Policies as Tax Rebate, Incentives, Energy cost concessions to customers	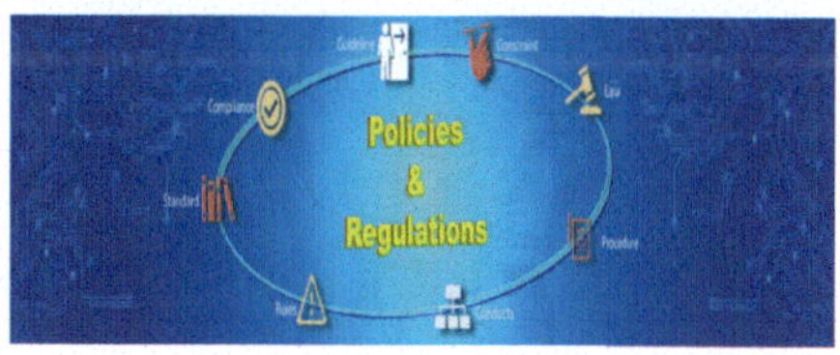
10. Net metering enabling customers to sell excess electricity to the grid, eliminating the need for on-site storage, and provides an incentive for installing renewable energy devices	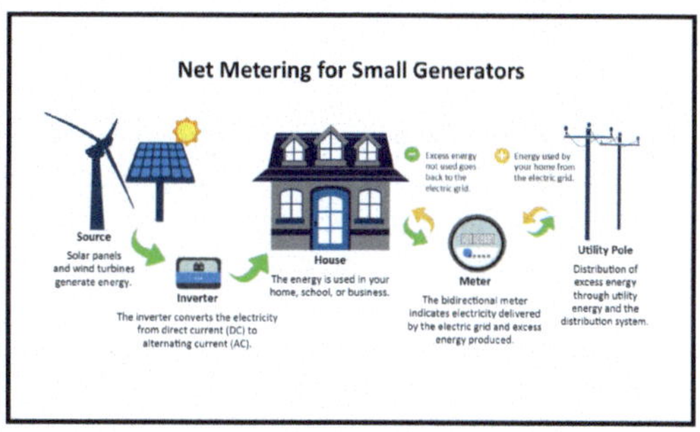

1.2.1 Renewable Energy Technologies—Taxonomy

Renewable energy technologies for electricity generation can be grouped [9] into:

1. **Dispatchable renewables**, such as hydro power, geothermal power and biomass power
2. **Non-Dispatchable renewables** that are also referred to as "variable" or "intermittent" renewables, such as wind power, solar photovoltaics, concentrating solar power,wave and tidal power and Waste-to-Energy Technology.

The dispatchability of an electricity generation source refers to the source's ability to be controlled in response to system requirements, such as variation in demand. In general, dispatchable renewables are constantly available (apart from maintenance needs) for production and offer high capacity factors2 (i.e. close to those obtained from fossil fuels or nuclear power plants, though with certain limitations). In contrast, electricity generation from non-dispatchable renewables depends on meteorological conditions. As a consequence, capacity factors are modest and grid operators cannot fully plan the electricity generation from these sources; only a fraction of the installed capacity can be considered as *statistically* dispatchable and an appropriate amount of back-up capacity is needed in power grids with a significant share of variable renewables.

1.3 How Pandemic is Changing the Energy Equation

1.3.1 Global Scenario

One of the driving forces of a country's economy is energy. The energy situation of a country takes various states depending on several factors, one set of factors which are governed by the people as industrial development and technology development leading to increase in standard of living, etc. and another set of factors due to nature as earthquake, floods and people health related events as the current COVID-19 pandemic [10]. The energy sector scenario during these untoward natural calamities takes a drastic fluctuation. There may be a rise or decrease in electric power consumption, due to increased activities as increasing the production of certain products, working more hours; or decrease in energy consumption due to shift in the area of usage as industry to domestic, change in life style or damage of electric sector physical resources.

Today, the growth and development in energy sector is thus interrupted due to the outbreak of Corona since the first quarter of 2020. Figure 1.9 depicts the shift in energy [11] source from conventional to RET due to Corona.

Social distancing has been advised to be the best solution to minimize the spreading of Covid-19, as it is a surprise novel pandemic and no medicine is available to cure or stop the spreading. Hence, most of the countries have imposed nation wise

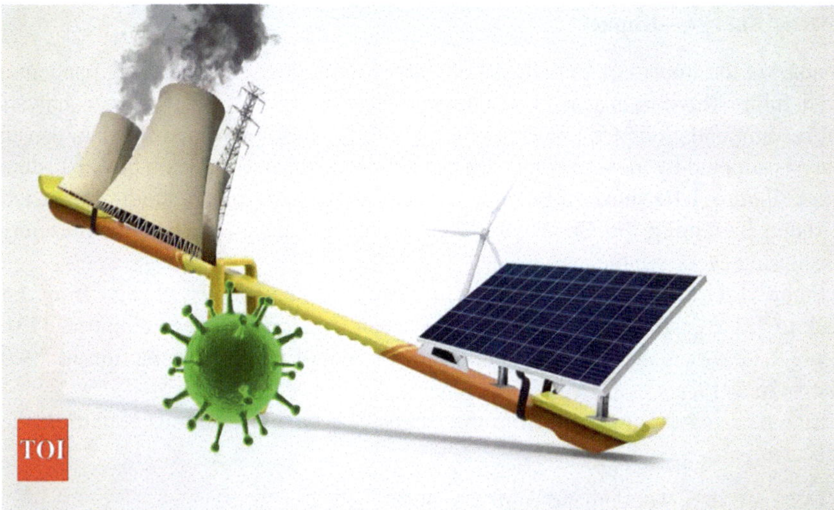

Fig. 1.9 Shift in energy source

strict lockdown. The lockdown has resulted in social restriction, travel ban, unemployment and working from home policy forcing most of the people to stay inside the house. This has affected the normal business operation and shifted the work from industries and offices to work from home. Industries moved to the minimum manual operation resulting in reduced electric power consumption. Business reduced their operation; travel ban almost collapse the transport industry, small business almost stopped, schools, universities moved to online mode and most other sectors adopted working from home policy. A large number of people of various cadres have lost their jobs, due to which they are not able to use their electrical gadgets, for want of money to pay their bills.

Social and Economic Impact:

The present COVID-19 pandemic event has impacted the social and economic state of a country in several areas; to mention the important ones:

- Global stock market crashed by more than 25% in March 2020.
- The international oil price has dropped to the lowest level since 2003 due to the combined effect of COVID-19 related demand drop and other business issues.
- Several schools, colleges and universities have shifted to online classes to continue education which also affected the power systems. Almost all offices have worked online from the residence of employees.
- Research and Development has completely shifted from various areas towards developing means and methods to manage COVID-19. Increase in manufacture of medical equipments and items as masks, PPE kits, sanitizers, Oxy-Meters, gloves, respirators, testing kits, medicines etc.

Electric Energy—Impact

To analyze the impact of COVID-19 on energy and power sector, during pandemic and in future, Ravikiram et al. [12] in their publication "COVID-19: "Impact analysis and recommendations for power sector operation", have presented the data driven analysis adopted by them to determine the impact of the pandemic on power system sector. Figure 1.10 shows the detail research framework of the study conducted by them, pertaining too Covid-19. The research has been conducted extensively considering every minute aspect.

Figure 1.11 Shows the Reduction in electricity demand (% change from 1st week to 3rd Week in March-2020) (a) European (b) Non-European countries [13]. Figure 1.12 shows the drastic reduction in All India Energy consumption in 2020 relative to 2019.

Summarizing the findings, from the extensive data about various countries, the facts are:

- Decrease in power demand from 5 to 40%
- Increased renewable generation

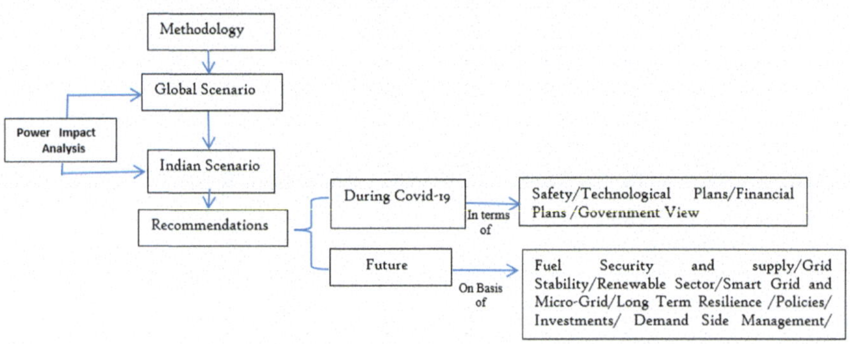

Fig. 1.10 Research framework of the study conducted by Ravikiram et al. Pertaining to Covid-19

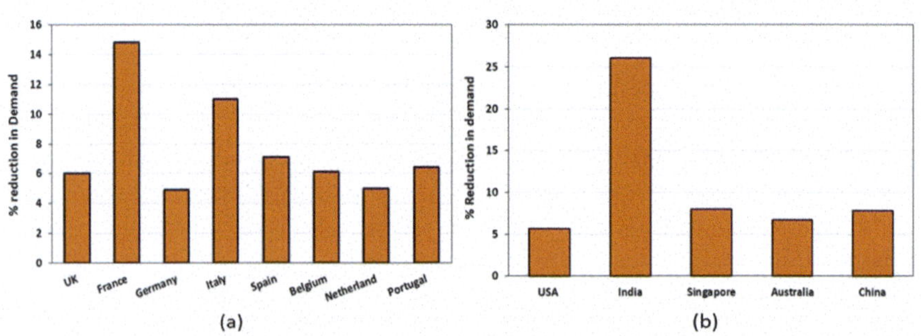

Fig. 1.11 Reduction in electricity demand (% change from 1st week to 3rd Week in March-2020) **a** European **b** Non-European countries

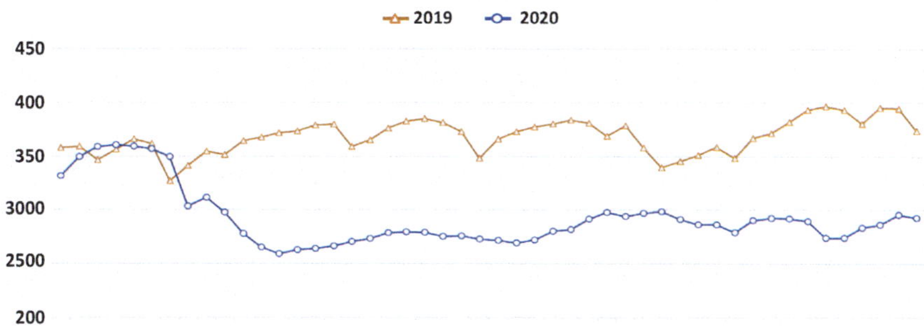

Fig. 1.12 All India energy consumption in GWh

- Decrease in industrial and commercial loads by 70%, and increase in residential loads
- Increase in bad debts for companies
- Economic drop to the government as government has granted relaxation for payment of utility bills.
- The mass reduction of demand patterns resulted to shut down some conventional generators based on fuel availability particularly thermal, gas and diesel power plants are operated economically.

Effect on Quality of Electric Power

In the wake of COVID-19, the technical challenges encountered by the electric power industry are:

- Reduced electric power consumption from grid effecting the reliable operation
- Supply disruptions due to delayed maintenance

1.4 Benefits to the Public, Society, Government and the Country

1.4.1 Introduction

All stake holders of Renewable Energy Technologies have their respective role in decision making and reaping the benefits of RET. Figure 1.13, depicts the taxonomy of the various electric energy stakeholders. The stakeholders of any service industry are the government, manufacturers of various related equipments and suppliers, and consumers. These stakeholders [14] are all people. In the context of RET, we have to consider a new group of stakeholder which is the element of the nature as air and environment.

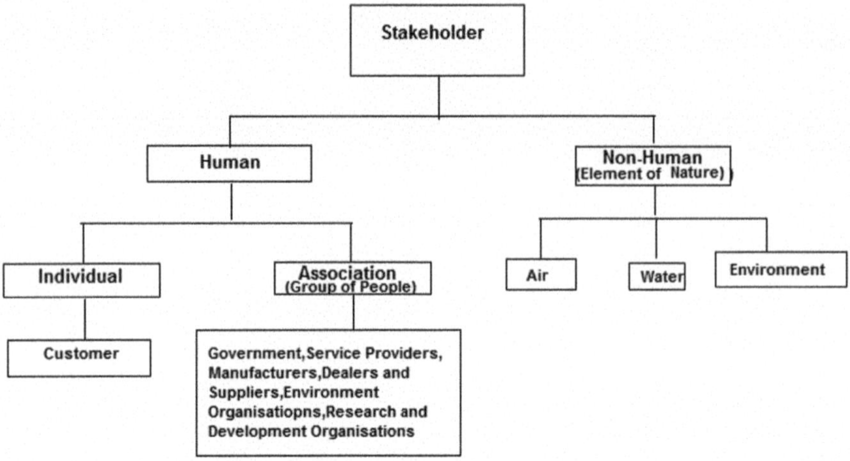

Fig. 1.13 Taxonomy of stakeholders

For successful implementation of any technology, the benefits expected by each stakeholder shall be analyzed and imbibed in the design. The design, development and implementation of RET shall be in support of benefitting all the stakeholders.

1.4.2 Analysis of Benefits

The steps for analysis of benefits are:

i. Identifying the benefits
ii. Determine the parameters to quantize the benefits
iii. Collect data pertaining to the parameters. Compute the statistical parameters as average, deviation etc., and obtain the trends. Draw conclusion and communicate results to draw information and take corrective measures.

1.4.2.1 Identifying the Benefits

The benefits to the various stake holders are:

- Inexhaustible Energy Source at the point of usage.
- Less global warming, Contributing towards cleaner and eco-friendly environment.
- Low Cost/Unit.
- Lesser dependence on imported and centralized energy.
- Increased Productivity improving air quality.
- Increasing energy system reliability.
- Support local economic development.

- Improving Public health.
- Negligible Transmission Losses.
- Add Value to Property.
- Job creation in various areas as manufacturing, installation, testing and operation and at various levels of the Management and Project hierarchy.
- Increases job opportunities for skilled personnel in various cadres and educational backgroundand un-skilled labour.
- Less maintenance cost.
- Improve energy security.
- Improve access to energy.
- Increased standard of living—proving facilities to increased use of electric gadgets thereby comforts.
- Increase Research and Development.
- Effective management of the waste generated at the consumer end, by adopting Waste-to-Energy practices.

1.4.2.2 Determine the Parameters to Quantize the Benefits

The parameters that enable quantize the benefits are:

1. Generation, transmission and distribution costs.
2. Generation output.
3. Unit cost of power to consumer.
4. Losses.
5. Power system down time.
6. Faults and failures.
7. Employment.
8. Imports in oil.
9. Carbon emissions.

1.4.2.3 Quantizing Benefits

Quantizing the extent of these benefits requires large amount of data collection which is not possible by individuals, therefore we have organisations as IRENA, MNRE,EEA,EIA etc who provide complete support from providing statistics, standards, schemes and training.The quantum of benefits over the next few decades predicted by IRENA [15] is given in Figure 1.14.

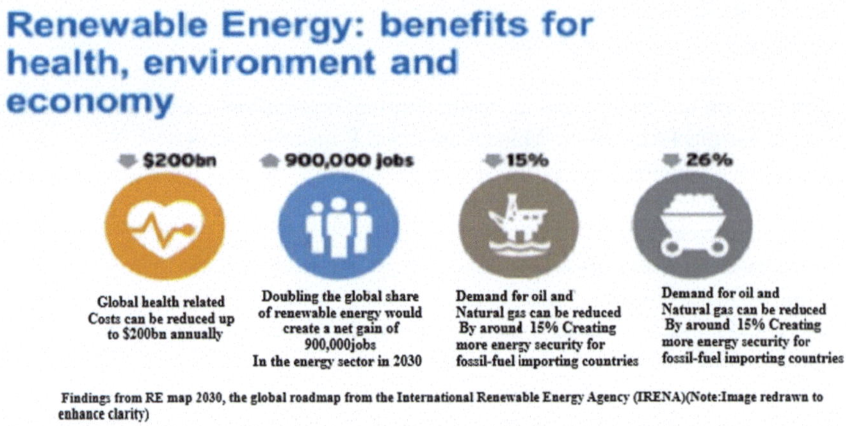

Fig. 1.14 Quantum of benefits predicted for next few decades

1.5 Environmental Impact of Renewable Energy Technologies

1.5.1 Introduction

There could be no process that would not result in certain by products apart from the final output. These by products can be categorized as:

i. Based on their physical state as: solid, liquid or gases
ii. Based on their energy content: as reusable or non-reusable. If they are reusable, they can be used for some other purpose else need to be discarded safely.
iii. Based on their effect: as hazardous or non-hazardous.

All the by products should be handled in accordance with standard norms, not causing any negative effect on people or environment. RETS' were conceived of not only to ease the burden on the grid but also to reduce the pollution of the environment by reduced emissions of greenhouse gases. It is also observed that the renewable energy projects are complex to install and are local environmental and condition sensitive.

The two main aspect of environment is air pollution, normally created by the discharged water from houses, industries, and polluted rain, discharge of used oils and liquids containing poisonous chemicals and heavy metals like mercury, lead, etc. And discharge of gases from industries, power generation units and vehicles. Carbon dioxide emission during generation of electric power by different energy resources given in Fig. 1.15, indicates the reduced emissions by non-conventional sources.

It is the responsibility of the public and government to take measures in reducing the environment pollution. The developers of the technology shall imbibe the features of not producing environment non-friendly waste products in any form as gas, liquid

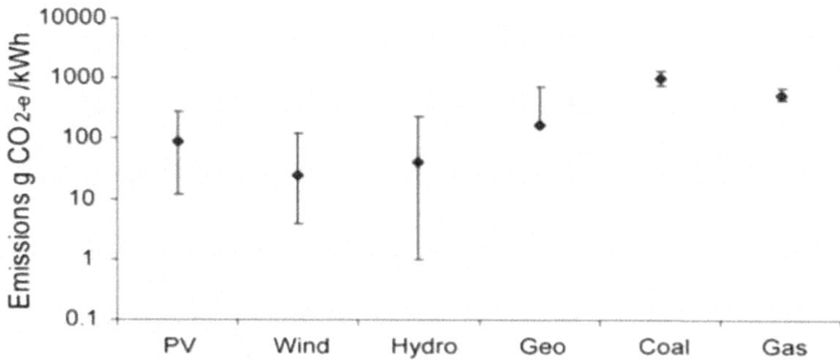

Fig. 1.15 Carbon dioxide equivalent emissions during power generation

or solid in the system design. The government of several countries therefore have come into a consensus and signed the crucial Paris Agreement [16] pertaining to environment protection. The Paris Agreement is presented in the next section for the knowledge and awareness of the beneficiaries of this report.

1.5.2 Paris Agreement

The Paris Agreement signed in 2016, within the United Nations Framework Convention on Climate Change (UNFCCC), deals with greenhouse-gas-emissions mitigation, adaptation, and finance.

The aim of the agreement is to decrease global warming, thereby significantly reduce the risks and impacts of climate change. The strategy involved energy and climate policy including the so-called 20/20/20 targets, namely the reduction of carbon dioxide (CO_2) emissions by 20%, the increase of renewable energy's market share to 20%, and a 20% increase in energy efficiency, depicted in Fig. 1.16.

The Paris deal is the world's first comprehensive climate agreement, under which each country must determine, plan, and regularly report on the contribution that it undertakes to mitigate global warming.

1.5.3 Facts—the Current Scenario Across the Globe

Almost every country has taken major measures to protect the environment, and taken steps, primarily establishing various governing bodies to encourage research and development, implement renewable energy technologies and providing financial schemes. Educating public and introducing a course in the curriculum at under-graduate and post-graduate levels.

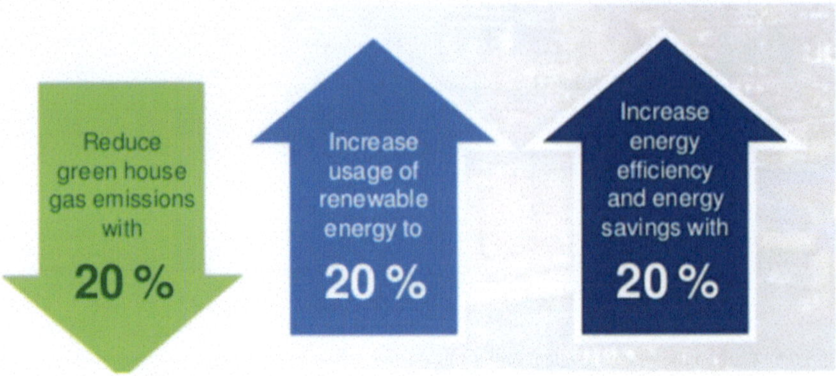

Fig. 1.16 20-20-20 policy

The initiatives taken by India and in Europe are presented, which almost represent that taken by various other governments with certain modifications suiting to their respective environment.

1.5.3.1 INDIA—India is Making Great Strides Towards Cleaner Energy

India is now at the fifth global position for overall installed renewable energy capacity. India has been taking measures to curtail environmental pollution since the 1980s, considering air, water, land and waste issues. Over the years, the government has been progressively strengthening rules and norms to combat air pollution. Government has adopted the National Clean Air Programme (NCAP), which focuses on monitoring of environment pollution levels and enforce the norms.

India is particularly vulnerable to climate change impacts. It is exposed to growing water stress, storms, floods and other extreme weather events. Adaptation and resilience of the energy system to these extreme climate conditions, is to be managed technically and socially.

The government of India has also made significant progress by way of encouraging the use of liquefied petroleum gas [17], and reduce the use of traditional biomass in cooking, the chief cause of air pollution, indoors and outdoors. India also continues to promote off-grid electrification solutions, including a shift toward using solar photovoltaic's (PV) for cooking and charging batteries.

Fig. 1.17 Renewable power generation in the European Union has nearly doubled since 2005. (Unsplash)

1.5.3.2 EUROPE—Ease on Environmental Burden by Renewable Energy

Europe could reduce environmental issue problems as reducing greenhouse gas emissions by shifting from fossil fuel-based electricity [18, 19] to renewable source generation. Figure 1.17 shows the installation of wind energy system in ocean.

The benefit to the environment is in the aspects of:

1. Soil acidification: Fig. 1.18
2. Eutrophication—Fig. 1.19 where freshwater becomes overloaded with nutrients, causing algal blooms and low oxygen levels—and the formation of particulate matter, a type of air pollution linked to 379,000 deaths in Europe in 2018.

Renewable energy sources are however non-zero impact sources. Producing power by incinerating waste can affect freshwater eco toxicity, while biomass energy is associated with intensified land occupation and the formation of particulate matter. Meeting EU emissions-cutting goals will require an even faster expansion of renewable sources, requiring a power sector based 70% on renewable by 2030.

Given this expected growth, the EU will need to tackle the potential environmental impacts of renewable, EEA said. The suggested methods are:

i. better reuse of materials could curb the environmental impact of mining the metals and purifying the silicon used to make solar PV panels.
ii. Adopting the 3R Principle of Waste-To-Energy Management

Fig. 1.18 Soil acidification

Fig. 1.19 Eutrophication

iii. Conducting Life Cycle assessment (LCA) studies: An LCA estimates resource requirements, energy use, and environmental impacts of products or services at all stages of life

Shift to renewable energy reduces key environmental burdens, EU says | Hindustan Times.

1.6 Role of the Government and Public in Implementing RET

Successful adoption and implementation of RET by the utilities requires, primarily educating and creating awareness to public regarding various renewable energy sources and technologies. The responsibility of educating the public rests with the Government.

As existing energy utilities play a key role in determining the adoption of RET, the utility structure and the strategy for integrating renewable should be reviewed and modified by the Government and other decision makers.

Renewable Energy Technology have the advantage that, they can be installed in small sizes by domestic consumers and in large sizes by big industries or organisations, suiting their required energy demand. Hence there is need to educate every single individual and organisations as industries, commercial establishments, educational institutions to mention a few.

The use of RET calls for awareness of its neccessity and sometimes a change of lifestyle. Electric Energy research and development and demonstration projects should be encouraged to improve information and raise public awareness. The media as radio, TV, news papers are the best means of spreading awareness. In the modern day with the sophisticated technology available—Video sessions can be recorded and circulated to various educational, business and industrial organisation. On-Line training sessions can be conducted. Financial incentives should be provided for manufacturing, installation, and import subsidies to the suppliers and concessions for consumption to the consumers.

The public shall take maximum advantage of all training programs, utilize the funds in implementing RET, reap the benefits thereof and transfer their knowledge and experience to the society so as to benefit one and all.

1.7 Various Initiatives by the Government—India

1.7.1 Introduction

Government is at the apex of the hierarchy of stakeholders for framing of any policy to implement renewable energy technologies. Ministries and Organizations have been constituted by Governments for RET, as RET itself has become a vast area by itself. Effort has been made to present the policies of India, in the present day context.

1.7.2 India Policy

January 10, 2021 2:08 PM Financial Express.

Growing the share of renewable energy in India's total energy mix will require additional policy oversight and ensuring that the interests of all stakeholders involved—developers, consumers, *DISCOM* (Distribution Company (In India)), and the government—are considered and protected.

Schemes under national solar mission framed by the government of india (GOI) are:

- Develop solar parks and ultra mega solar power projects
- Installation of 300 MW of grid-connected solar PV power projects by defence establishments under Ministry of Defence and Para Military Forces.
- Pilot projects for development of grid connected solar PV power plants on canal banks and tops.
- Generation based incentives programme for small solar power generation
- Grid connected rooftop solar programme
- Pradhan Mantri Kisan Urja Suraksha Evam Utthaan Mahabhiyaan (PMKUSUM) scheme
- **YOJANA (AJAY): PHASE-II**: Distribution of 70 lakh solar study lamps, off-grid and decentralized concentrated solar thermal (cst) technologies for community cooking, process heat and space heating & cooling applications in industrial, institutional and commercial establishments' scheme.
- Issuance of duty exemption certificates for manufacturing of wind turbines
- **Waste to Energy** programme: It is aimed at generation of biogas, BioCNG and Power from different wastes, such as vegetable and other market wastes, slaughter-house waste, agricultural residues and industrial wastes & effluents. In addition to Bio-CNG/Biogas, biogas plants generate organic fertilizer as a by-product which is valuable for agricultural fields.

Figure 1.20 shows South India's biggest plant at Jawaharnagar, Hyderabad dump yard. This plant generates 63 MW of power through municipal solid waste, and Fig. 1.21 is the Waste-To-Energy Process.

Fig. 1.20 South India's biggest plant at Jawaharnagar dump yard

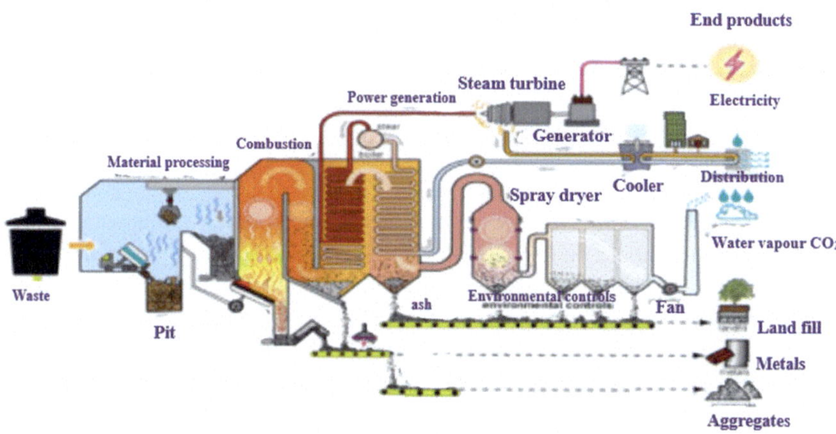

Fig. 1.21 Waste-to-energy process plant

1.7.3 Scheme to Support Promotion of Biomass Based Cogeneration in Sugar Mills and Other Industries Other Than Sugar Mills (Upto March 2020) Off-Grid

The key points are:

- A new scheme to support promotion of biomass based co-generation in sugar mills and other industries (effective up to March 2020) has been notified on 11.05.2018, which would enable meet the need of electrical and thermal energy of many industries for their operations. Biomass Under the scheme a Central Financial Assistance (CFA) @ Rs. 50 Lakhs per MW on installed capacity will be provided.

1.7.4 *Off-Grid Renewable Power - Biogas Power (Off-Grid) Generation and Thermal Application Programme (Bpgtp)*

BPGTP is implemented through the Agriculture and Rural Development Departments of the States, Dairy Co-operatives, State Nodal Agencies (SNAs), Biogas Development and Training Centres (BDTCs), Khadi and Village Industries Commission (KVIC) and National Dairy Development Board (NDDB) from the year, 2018-19.CFA for BPGTP varies from Rs.25,000/- per kW to Rs.40,000/- per kW for power generation as per the generation capacity slab and Rs.12,500/- per kW to Rs.20,000/ - per kWe for thermal applications respectively. The CFA rates also vary depending upon the category of beneficiaries and regions/ States.

1.8 Standards for Renewable Energy Technology

1.8.1 *Introduction—Need for Standards*

Standards specify a specification for any system to comply with, without which it will not be know what should be the various parameters to which the system should be manufactured, assembled, tested and installed. It is known that the systems are made up of modules, so without standards it is not possible to interface them. Also, the modules will be made by several manufacturers, and deployed to various locations globally. Standards are the preliminary planning document along with the customer requirement specification.

Governments all over the world have initiated several programmes, to encourage introduction of new technologies and also expansion of existing technologies. Therefore, there is need to develop standards to embed the new technology modules in the existing systems.

Every country has its own standards, but however they ensure that the systems designed to their standards can be interfaced with the systems of other countries. Reproducing the standards of all the countries would be extensive in length, so a few are discussed, which enabled to get a clear understanding and importance about the application of standards.

1.9 Financial Aspects of Renewable Energy Technologies

Finance is one of the main major resource for any project. The planning for finance starts from the conceptual of the project. Finance is also one of the barriers for sustainable development. Financial planning is determining the cost of the various heads as land, machinery, material and manpower; and also identifying the sources of the finance as banks and various Financial Institutions.

Renewable Energy Project (REP) is high capital intensive, hence it is required to be abreast with the laws, policies, regulations, and incentives offered by the Governments that could better facilitate the easy and fast implementation of the project. Renewable energy projects are still viewed as higher risk investments, requiring efficiency improvement.

Setting the cost of the REP requires determining the market structure in any country. Unlike the conventional energy sector, the onus of developing the renewable energy sector is with the private sector. The challenge for the sector is then to mobilize private capital at a pace and rate in tune with the policy targets and goals. It is built on the premise that private sector has the potential to fund the sector; however, it requires an enabling framework to be created by the government.

The opportunities and avenues available in India is presented in this text at Sect. 1.9.1. This would give an insight of the strategic approach towards undertaking the REP, be it anywhere in the country.

1.9.1 India's Energy

The International Energy Agency (IEA) in partnership with National Institution for Transforming India, NITI Aayog, in 2020, Fig. 1.22 released the review of India's energy policies. They provide recommendations supporting the government's objective to promote effective energy markets and boost deployment of RET. As per the report of IEA, the country's demand for energy is set to double by 2040.

The important key aspects of the review report that indicate the rise of India in global energy markets are:

- India has made impressive progress in increasing citizens' access to electricity.
- The Government of India has also made significant progress in reducing the use of traditional biomass in cooking.
- **Major energy reforms lead to greater efficiency, therefore increased** access to affordable energy has raised the living standards of all segments of the population.
- India now has the institutional framework to attract more investment for its growing energy needs.
- The IEA welcomes the government's decisions to allow private-sector investment in coal mining, and to open up the country's oil and gas retail markets.
- India is making energy security a priority.

The new IEA report on India's energy policies was launched at NITI Aayog's headquarters in New Delhi

Fig. 1.22 Meeting at NITI Aayog

- India has made important progress towards meeting the United Nations Sustainable Development Goals.
- Energy technology and innovation enables "Make in India" manufacturing initiative.

1.9.2 India—Schemes

Initiatives for Loans and International Funding [20]

- Reserve Bank of India has included renewable energy projects under priority sector category for granting of loans to Institutions and individuals.
- Department of Financial Services (DFS) has advised all Public Sector Banks to provide loans for grid connected rooftop solar systems.
- Multilateral grant of several million dollars from World Bank (WB) has been approved for solar rooftop programme.
- Indian Renewable Energy Development Agency (IREDA) has formulated a scheme of low-cost financing.

- Demand aggregation exercise for government buildings of various States has been initiated for installation of rooftop solar projects under technical assistance programme.

1.10 Modern Power Electronics Technologies

It is not just enough to discuss about increasing the quantity of electric energy generation, by identifying various alternative energy sources to meet the rising customer demands. Customers electrical gadgets have become more sophisticated demanding the need for reliable and intermittent high quality power supply. There is need to develop equipments and techniques to continuously monitor the electric system performance and take immediate corrective action. The equipment's and techniques should also enable collection, storage and analysis of the vast data collected through various sensors and measuring instruments, so as to enable predictive studies. The advancement in technology in the areas of power electronics, electric machines and drives, power systems and renewable, instrumentation, computer based systems as microprocessors, microcontrollers, information technology integration systems as SCADA is making it possible to make several advancements in electric grid leading to meeting the goals of power in terms of quantity and quality. All these technologies when integrated lead to generate and deliver quality power.

In the context of RET the two major areas demanding research and development are

a. Integration of Renewable Energy Source to grid and customer power supply system, as this aspect leads to many losses and quality problems.
b. Since Renewable Energy is not available 24/7 but exceeds the customer demand when available there is need to have storage systems.

1.10.1 Quality Parameters

The technical parameters that govern the quality of power are:

i. voltage quality—interruptions, variations, unbalances, flicker, sags, and swells
ii. Frequency
iii. Harmonics
iv. Power Factor
v. Losses
vi. Ripples

Today's electronic loads are susceptible to transients, sags, swells, harmonics, momentary interruptions, and other disturbances which are a cause for concern. Figure 1.23 shows the voltage as per IEEE–P1159—Recommended Practice for

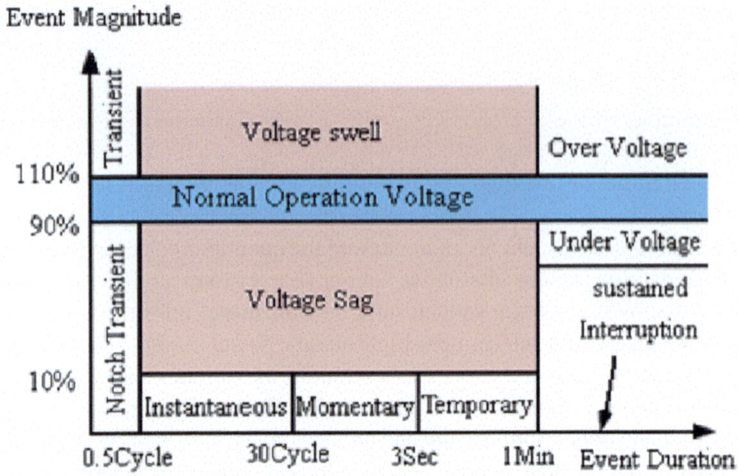

Fig. 1.23 Voltage reduction standard of IEEE Std. 1159–1995

Electric Power Quality Monitoring. Use of RET as a standalone power source or integrating with the grid requires managing the power quality parameters.

1.10.2 Integration of Renewable Energy Source to Grid and Customer Power Supply System

In the present day, several technological components and concepts are available to integrate RET into the power system grid. They are:

- Flexible generation to accommodate the advances in the hardware and software technologies, as the power grid is no more a pure electrical hardware system but has become and intelligent system.
- Grid extension is the connection of the consumer to central grid system
- Smart grid technologies
- Storage technologies
- Advances in wind and solar PV technologies allow them to be used over a wider range of conditions.
- Demand Side Management (DSM) controls
- Advanced Metering Infrastructure (AMI)
- Smart inverters
- Phasor Measurement Unit (PMU)
- Fault Ride Through (FRT) capabilities
- Mini-Grids and Islanding systems

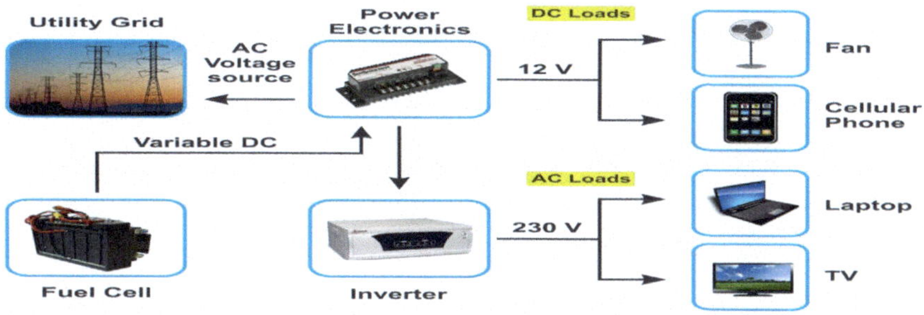

Fig. 1.24 Power electronics as a key element

Figure 1.24 shows a general schematic of the modern power electronics component interfacing the renewable energy source system and the consumer loads. It can be seen that the DC output of the RET can be used directly for DC loads or can be converted to required level of AC by Inverters to meet the demands of AC loads.

The integration of RET into power grids requires substantial transformations to increase the flexibility of the existing grids, in meeting the following requirements and conditions:

(a) To allow to-and-fro of electricity flow, between centralized power plants and users i.e. grid and consumer
(b) To improve grid interconnection at the regional and international level aimed at addressing increasing balancing capabilities, flexibility, stability and security of supply; and
(c) To introduce energy storage capacity to store electricity (energy) from variable renewables generation when production exceeds demand.

1.10.3 Storage

The rapid growth in renewable power has driven demand for electrical energy storage systems (ESS). The various energy storage technologies currently available are depicted in Fig. 1.25 [21]. Each energy storage technology has its own implementation requirements, therefore the process of selecting an ESS requires a multi-level understanding of the components involved.

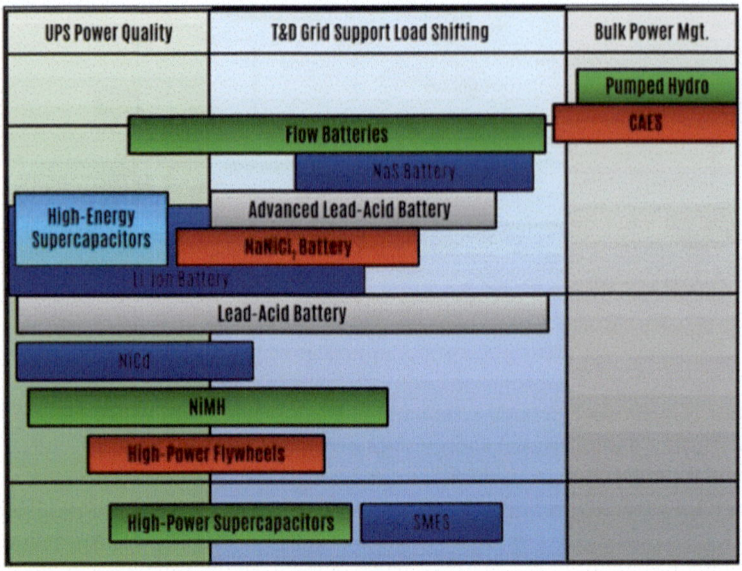

Fig. 1.25 Available technologies for energy storage

1.11 Opportunities and Avenues for Youth in Renewable Energy Technology from Research, Development to Implementation Areas

1.11.1 Introduction

On the introduction of any new technology and policy the technical, financial, social and political feasibility shall be considered. It is the social responsibility of the Government, Organisations, Institutions and all individuals of the Nation, to realise the need and ensure that any technology introduction would not only give financial benefits, fulfil social need requirement but also maximize employment opportunities.

Renewable energy industry is more labour intensive, compared to conventional energy industry. The number of jobs created for each unit of electricity generated from renewable sources is more than that for fossil fuels.

1.11.2 Key Facts of Employment

The statistics provided by International Renewable Energy Agency (IRENA), reveals the rise in employment in the solar, wind, bio-energy, and other energy generation technology areas over the years, as compared to employment in hydro-power generation area. This is a base for encouragement to youth and society to embrace these new technologies, and indication to educational institutions that there is need to increase the imparting of knowledge in the latest technologies of power generation.

Assessment of RET by IRENA, revealed the facts that solar photovoltaic (PV), bio energy, hydropower and wind power industries have been the biggest employers. Figure 1.26 shows the evolution of IRENA's renewable energy employment estimates in million jobs since 2012 to 2019. Table 1.3 depicts the percentage employment technology-wise, and the comparative graph in Fig. 1.27.

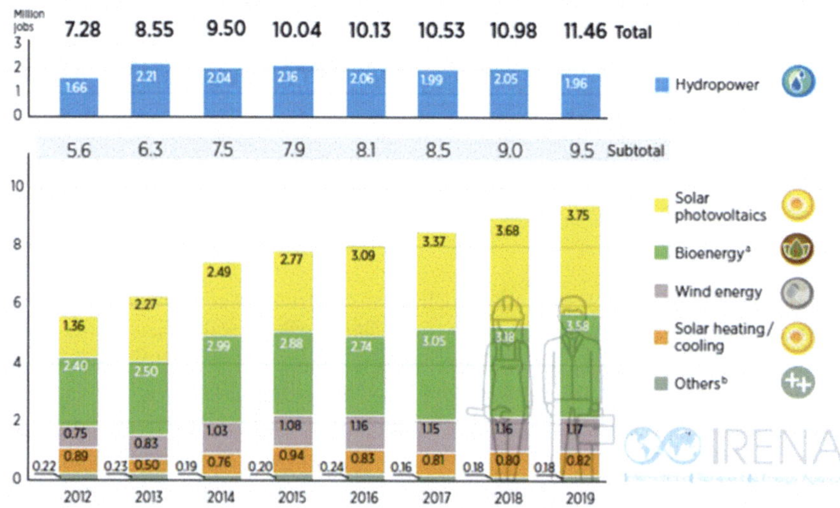

Fig. 1.26 Renewable energy employment (million jobs) 2012–2019

Table 1.3 Percentage employment technology-wise

YEAR	Hydro-power	Solar photovoltaics	Bioenergy	Wind energy	Solar heating/ cooling	Others
2012	23	19	33	10	12	3
2013	26	26	29	10	6	3
2014	22	26	31	11	8	2
2015	21	28	29	11	9	2
2016	20	31	27	12	8	2
2017	19	32	28	11	8	2
2018	18	34	29	10	7	2
2019	17	33	31	10	7	2

Note Computed from Fig. 1.22

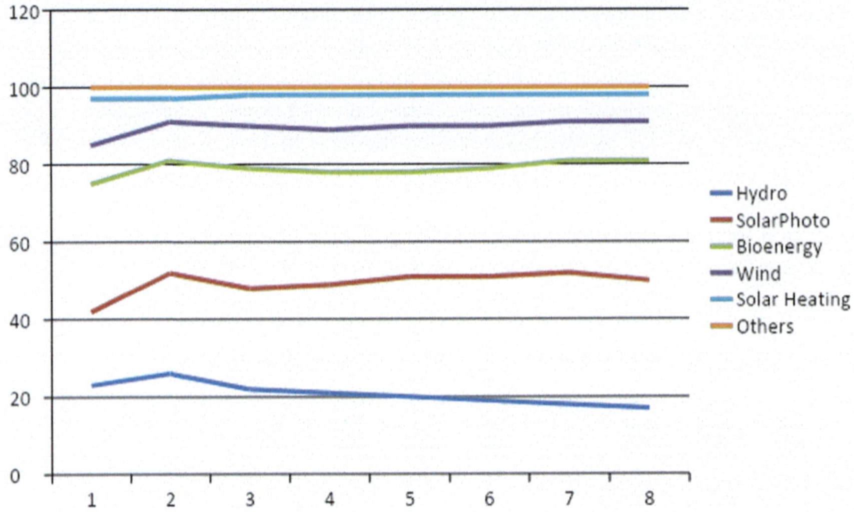

Fig. 1.27 Percentage job trend technology—wise 2012–2019

Renewable energy, now predominant in new electric power capacity [IRENA], has proven especially flexible, cost-effective, and resilient in the face of the 2020 health and economic crisis. Lockdown measures have significantly changed the power mix. Daily data collected for more than 30 countries (Source IEA), which represent over one-third of global electricity demand, show that the extent of demand decline depends on the duration and stringency of lockdowns. A month full lockdown reduced demand by 20% on average, or over 1.5% on an annual basis. Figure 1.28 depicts a view of the scenario.

Longer lockdowns, slower economic recovery, and wide diffusion of Covid-19 in developing countries could cut demand even further. The Power System Operation

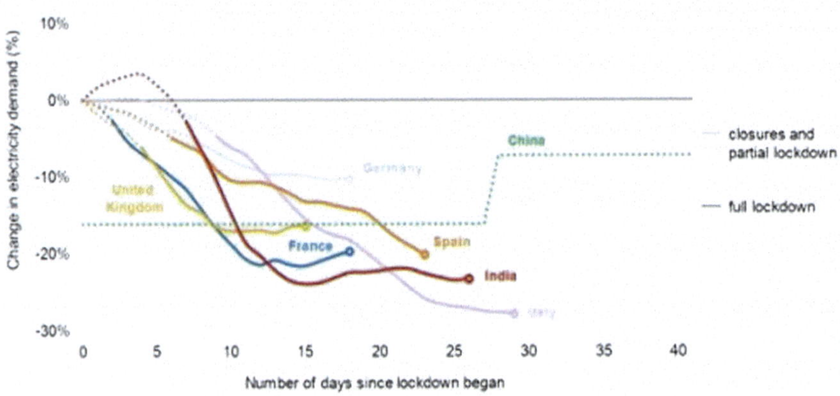

Fig. 1.28 *Source* Global Energy Review 2020

Corporation of India reported that electricity production from coal—India's primary source of electricity—fell by 32.2% to 1.91 billion units (kilowatt-hours) per day, compared to the 2019 levels.

The post-COVID agenda put forward by the International Renewable Energy Agency (IRENA) proposed to create some 5.5 million transition-related jobs over the next three years, paving the way for longer-term development.

With a number of renewable energy manufacturers looking to shift base from China to diversify their supply chains in the wake of the Covid-19 pandemic, MNRE [has urged states to provide incentives for setting up designated manufacturing hubs in India.

The proposal has attracted interest from Madhya Pradesh, Odisha and the Tuticorin Port Trust in Tamil Nadu to set up designated equipment manufacturing parks in the country.

MNRE has asked the governments to identify land parcels of 50–500 acres for setting up the parks, which shall cater to both domestic and international needs. These hubs will manufacture equipment such as silicon ingots and wafers, solar cells and modules, wind equipment and other ancillary items such as back sheet, glass, steel frames, inverters, and batteries.

The report from the International Renewable Energy Agency (IRENA) gives practical advice on key investment and policy decisions for the crucial post-COVID recovery. It is rooted in the comprehensive long-term energy transformation strategy provided by IRENA's first *Global Renewables Outlook*.

Among other findings:

- Post-COVID recovery measures could drive a lasting shift in the global energy mix.
- Linking the short-term recovery to medium and long-term strategies is important in achieving Sustainable Development Goals (SDGs) and meeting the Paris Agreement.
- Energy transition investment can boost GDP and create jobs in the 2021–23 recovery phase.
- Green investment will be vital to mobilise finance for the transition.
- An accelerated energy transition could add 5.5 million more jobs by 2023.

Key policy recommendations for 2021-23 recovery phase are:

- Raising national targets and enhancing climate pledges.
- Securing strategic finance, investing in transition-related infrastructure.
- Safeguarding existing renewable projects and ensuring stable policy frameworks for transition-related infrastructure.
- Creating renewable-based, efficient infrastructure, incentivising electric vehicle and biofuel use, and promoting behavioural changes to reduce the impact of travel.
- Co-ordinating with industries and developing education and training to minimise skills gaps.
- Diversifying supply chains, developing local industries, providing finance and promoting supplier development programmes and industry clusters.

- Maintaining energy access initiatives, ensuring reliable power supply and supporting distributed renewable energy solutions.

1.12 Metrics and Measures of Electrical Energy

1.12.1 Introduction

Decision-making procedures backed by hard numbers is a great way for renewable energy developers to optimize their performance. Key performance indicators (KPIs) can be used to help managers quantify how their renewable energy business performs to make smart decisions. Isolating time wasting tasks, predicting financial feasibility, comparing asset performance and many other metrics can be used to make businesses run smarter.

There are an infinite number of KPIs which vary according to industry and business needs. Here's a few important metrics for developers and operators of every kind of renewable asset: wind, solar, biomass, hydro, etc.

The performance measures can be classified into two main categories:

1. Technical.
2. Financial or Economic factors

Technical Factors:

Before we delve into the technical factors we need to know the terms related to the specification of a power supply source which could be a generation unit or a sub-station. The terms are:

1. **Connected Load**

Connected Load is the sum of continuous ratings of all the equipment connected to the supply system. Each consumer has certain equipment installed at his premises. The sum of the continuous ratings of all the equipment in the consumer's premises is the "**connected load**" of the consumer.

For example, if a consumer has connections of five 100-W lamps and a power point of 500 watts, then the connected load of the consumer is $5 \times 100 + 500 = 1000$ watts. The sum of the connected loads of all the consumers is the connected load to the power station.

2. **Maximum Demand**

The Maximum of all demands (loads) on a power station during a given period is known as **Maximum Demand**. Generally, all consumers never switch ON all the devices at full load, therefore Maximum Demand is always less than or equal to the connected load. Maximum demand helps in determining the size and cost of the installation.

3. **Average Load**

The average of loads occurring on the power station in a given period, which could be a day, month or year is known as the **average load or average demand** given by Eqs. (1.3)–(1.5) respectively.

$$\text{Daily Average Load} = \frac{\text{Energy supplied (in kWh) in a day}}{\text{Hours in a day(24 hrs)}} \tag{1.3}$$

$$\text{Monthly Average Load} = \frac{\text{Energy supplied (in kWh) in a month}}{\text{Hours in a month}(30 \times 24 = 720\text{ hrs})} \tag{1.4}$$

$$\text{Yearly Average Load} = \frac{\text{Energy supplied (in kWh) in a year}}{\text{Hours in a day}(365 \times 24 = 720\text{ hrs})} \tag{1.5}$$

Factors

1. **Capacity factor**, Eq. (1.6) is the ratio of actual electricity produced by a power plant in a year to the electricity that the plant could *theoretically* produce if operated continuously at full power during the same period.

$$Capacity\,factor = \frac{(actual\ energy\ produced\ or\ supplied\ in\ time\ T)}{maximum\ plant\ rating \times T} \tag{1.6}$$

Capacity factor is used in generation studies.

2. **Load Factor:** Load Factor. Eq. (1.7) is Defined as the Ratio of the Average Load Over a Given Period to the Maximum Demand (Peak Load) Occurring in that Period.

$$Load\,factor = \frac{average\ load \times T}{peak\ load \times T} \tag{1.7}$$

For determining load factor, the information required is:

1. Actual kilowatt hours used (kWh)
2. Peak kilowatt demand (kW)
3. Number of days

For Example:

Let total kWh = 30,000KWh.

Demand = 80 kW.

The number of days = 30 days.

Hours per day = 24 h.

Solution

Monthly Load Factor = (30,000/(80*30*24)) = 0.52.

If the load factor is high (above 0.50), it shows that the power usage is relatively constant; if it is low, it means a high demand is set. For improving load factor, electrical load operating at the peak time of the day should be shifted to non-peaks time.

3. **Demand Factor (DF)**

The demand factor, Eq. (1.8) is defined as the ratio of maximum demand to connected load.

$$Demand\ Factor(DF) = \frac{Maximum\ Demand}{Connected\ Load} \tag{1.8}$$

The value of the demand factor is usually less than one (i.e. DF < 1) as the maximum demand on the power station is generally less than the connected load. As both the numerator and denominator are in the same units W or kW Demand factor has no unit.

For Example: If the maximum demand on the power station is 100 MW and the connected load is 120 MW, then the demand factor = 100/120 = 0·83. Demand factor is vital in determining the capacity of the plant equipment.

4. **Diversity Factor (DvF)**

It is defined as the ratio of the sum of the individual demands of all consumers to the maximum demand of the power station. The diversity factor formula, Eq. (1.9) is as follows:

$$\text{Diversity Factor(DvF)} = \frac{\text{Sum of individual Maximum Demands}}{Maximum\ Demandof\ Power\ Station} \tag{1.9}$$

The sum of individual maximum demands is always greater than the maximum demand of the power station, therefore the diversity factor is always greater than unity one (i.e. DvF > 1). The diversity factor reduces the capital cost of the station and the rate of generation of electricity.

5. **Coincidence Factor**

Coincidence factor, Eq. (1.10) is defined as the reciprocal of the diversity factor, therefore is always less than unity.

$$\text{Coincidence Factor} = \frac{1}{Diversity\ Factor} \tag{1.10}$$

Financial/economic Measures:

The most important measures are:

1. **Return On Investment (ROI)**
2. **Net Present Value (NPV)**
3. **Internal Rate of Return (IRR)**
4. **Plant Availability**
5. **Performance Ratio (PR)**
6. **Capacity Utilisation Factor (CUF)**
7. **Greenhouse Gas Equivalency**
8. **Levelised cost of electricity (LCOE)**
9. **Energy Return On Energy Invested (EROI):**

1.12.2 Return on Investment (ROI)

The Return on Investment (**ROI**) is a simple calculation that compares a project's net profit to its original investment. In pre-development projects, the **ROI** is a critical measure that can be used to help decide if a project proceeds or not. It can also be used as a comparison tool, for example, an investor could choose between two projects to finance based on their projected ROI. Re-evaluating the ROI is an exercise that should be done periodically during the life of the project to ensure the project continues to be profitable.

Return on investment (ROI) measures the gain or loss generated on an investment relative to the amount of money invested. ROI is usually expressed as a percentage and is typically used for personal financial decisions, to compare a company's profitability or to compare the efficiency of different investments.

The return on investment formula, Eq. (1.11) is:

$$\mathrm{ROI} = (\mathrm{Net\ Profit/\ Cost\ of Investment})\mathrm{x}100 \tag{1.11}$$

1.12.3 Net Present Value (NPV)

A hundred years ago, you could buy a loaf of bread for a nickel, but now bread costs a few dollars. As time passes, inflation causes money to lose value. Cash that you have now is more valuable than the same sum of cash in the future. This concept is called the **time value of money**.

Net Present Value (NPV), Eq. (1.12) is a metric that is designed to simplify a project's financial model and account for the time value of money. By estimating yearly total cash flows in a financial model and assuming a discount rate (the rate at which money loses value over time), the project's total value, in today's dollars, can be calculated.

Interpreting NPV is straightforward. The higher the **NPV**, the more valuable the project. This metric can be really helpful in determining which project is more likely to **yield larger returns**.

The NPV formula given by equation is somewhat complicated because it adds up all of the future cash flows from an investment, discounts them by the discount rate, and subtracts the initial investment.

$$\mathrm{NPV} = \sum_{t-1}^{T} \frac{C_t}{(1+r)^t} - C_o \tag{1.12}$$

where,

- Ct = net cash inflow for the period
- CO = initial investment
- r = discount rate
- t = number of periods

1.12.4 Internal Rate of Return (IRR)

The Internal Rate of Return (IRR) is a metric in the same family as ROI and NPV. The ROI is a value that tells you how much your project returns, without considering the time value of money. NPV is one lump sum that considers the time value of money, but doesn't talk about returns.

IRR addresses these shortcomings by calculating an average yearly return while also considering the time value of money. In technical terms, IRR calculates the discount rate that makes the NPV of a project 0. This value provides a more accurate metric to evaluate a project's return than ROI.

The IRR formula, Eq. (1.13) is calculated by equating the sum of the present value of future cash flow less the initial investment to zero. Since we are dealing with an unknown variable, this is a bit of an algebraic equation. Here's what it looks like:

$$\mathrm{IRR} = \left(\frac{\text{Cash Flows Year1}}{(1+\mathrm{RRR})} + \frac{\text{Cash Flows Year2}}{(1+\mathrm{RRR})^2} + \cdots \frac{\text{Cash Flows Yeart}}{(1+\mathrm{RRR})^t} \right) - \text{Initial Investment} = 0 \tag{1.13}$$

As you can see, the only variable in the internal rate of return equation that management won't know is the IRR. They will know how much capital is required to start the project and they will have a reasonable estimate of the future income of the investment. This means we will have solve for the discount rate that will make the NPV equal to zero.

1.12.5 Performance Ratio (PR)

Performance ratio (PR), Eq. (1.14) is the factor through which the plant performance can be assessed. The **Performance Ratio** (PR) is an important metric for evaluating the efficiency of various **electricity generating facilities**. The PR, is the ratio between actual electricity production and theoretical production during a reference period expressed as percentage. The difference between theoretical and actual production is the losses caused by the different equipment installation.

The PR gives a measure of the quality of a given plant and is independent of location. It is also described as a quality factor.

The higher the PR value determined of a plant, the more efficient is the plant. High-performance plants can reach a performance ratio of near 80% losses.

There are two basic parameters needed calculate the performance ratio:

i. Y_f (Final PV system yield) and
ii. Yr (Reference yield).
iii. PR is the Ratio of Y_f and Yr.

$$\text{Performance ratio} = Y_f/Yr \tag{1.14}$$

where,

$Y_f = E/P_0$.

E = Net energy Output in KWh.

P_0 = Installed PV array in KW.

Yr = H/G.

H = Total plane irradiance in KWh/m^2

G = PV reference irradiance in KW/m^2

Performance Ratio (PR) of a plant for a period of time = Energy measured (kWh)/(Irradiance (kWh/m2) on the panel x Active area of PV module (m2) x PV module efficiency).

1.12.6 Plant Availability

This metric is a percentage that represents the relative time an asset is operating and generating value. A **solar power plant** with an availability metric of 90% for example is running the majority of the time, but 10% of the time it is offline. Comparing the availability of similar plants can be used to identify weak designs, equipment that is prone to failures or substandard maintenance practices. Armed with this knowledge, operators can target facilities to make performance improvements.

1.12.7 *Capacity Utilisation Factor (CUF)*

The capacity factor, Eq. (1.15) of a PV power plant (usually expressed as a percentage) is the ratio of the actual output over a period of one year and its output if it had operated at nominal power the entire year

$$\text{Capacity Utilisation Factor(CUF)} = \text{Energy measured (kWh)} / (365 * 24 * \text{installed capacity of the plant}). \quad (1.15)$$

1.12.8 *Greenhouse Gas Equivalency*

Developing new clean energy generation facilities, electricity can be produced without harmful **greenhouse gasses (GHG)**. GHG emissions are measured in terms of grams of carbon dioxide per unit per kilowatt hour (gm of CO2e/kWh).

The greenhouse gas equivalency calculation enables to determine how much of an impact the generation system has on the environment. This metric compares how many tons of $CO_2$is emitted by the clean facility.

Greenhouse Gas Equivalency metric can be used for many purposes as:

- Determine the quality of energy generated
- Compare various generation systems
- Can be used to evaluate the success of government incentive programs for green energy technology.

Life-cycle assessment METHOD CAN BE USED TO DETERMINE THE GREEN HOUSE GASES. Study by Intergovernmental Panel on Climate Change (*IPC*) [22] in its synthesis report (AR5) has published the GHG emissions of various energy sources as:

- Hydropower: about 4 g CO2e/kWh
- Wind power: about11 g CO2e/kWh
- Nuclear power: about12 g CO2e/kWh
- Solar power: around 41 g CO2e/kWh
- Natural gas: 290–930 g CO2e/kWh
- Oil: 510–1170 g CO2e/kWh
- Coal: 740–1689 g CO2e/kWh.

1.12.9 Levelised cost of electricity (LCOE):

LCOE, Eq. (1.16) is used to calculate the average cost per unit of electricity generated. It allows for the recovery of all costs over the lifetime of the plant, including capital, operation and maintenance, fuel (if any).

$$\mathrm{LCOE} = \frac{\sum(1_t + M_t + F_t)}{\sum \frac{E_t}{(1+r)^r}} \tag{1.16}$$

Simplified LCOE Calculation.

I_t = Investment expenditures in year t (including financing).

Mt = Operations and maintenance expenditures in year t.

F_t = Fuel expenditures in year t.

E_t = Electricity generation in year t.

r = Discount rate.

n = Life of the system.

We can compare different technologiesas wind, solar, natural gas, of unequal life spans, project size, different capital cost, risk, return, and capacities.

1.12.10 Energy Return On Energy Invested (EROI):

EROI, Eq. (1.17) is the ratio of the energy delivered by a process to the energy used directly and indirectly in that process. EROI is part of lifecycle analysis (LCA). Therefore,

$$EROI = \text{Quantity of Energy Supplied /Quantity of Energy Used in the Supply Process} \tag{1.17}$$

The units are usually in BTUs, kWhs or other units available to both energy supplied and used.

***Quantity of Energy Used in the Supply Process* includes:**

- Drilling, refining, construction, installation, smelting, operations, maintenance, decommissioning, transportation, roads, manufacture of specialized equipment and chemicals required for extraction

***Quantity of Energy Supplied* includes:**

- Electricity, useable heat, power for useful work

An EROI value of about 7 is considered break-even for developed countries. The US average EROI is about 40.

1.13 Peep into the Future

1.13.1 Hydrogen Energy Mission

Presenting the Union Budget 2021–22, the Union Minister of India for Finance Nirmala Sitharaman had announced that the Hydrogen Energy Mission will be launched in 2021–22, and India would join the race for producing the next big energy source. Green hydrogen energy is vital for India to meet its Nationally Determined Contributions and goals.

Hydrogen energy source is seen as the next big thing as its usage would lead to zero emissions. Figure 1.29 is a hydrogen fuel-powered bus in Seoul, South Korea. Green hydrogen, produced with renewable electricity, is projected to grow rapidly in the coming years. Hydrogen from renewable power is technically viable today and economically. However, green hydrogen energy technologies come with their own problems.

Fig. 1.29 This is a hydrogen fuel-powered bus in Seoul, South Korea

1.13.2 India's First-Ever Geothermal Field Development Project (PTI February 08, 2021, 07:29 IST)

The Union Territory Administration Ladakh, Ladakh Autonomous Hill Development Council (LAHDC)-Leh and Oil and Natural Gas Corporation (ONGC) Energy Centre, have signed a pact for establishing India's first-ever geothermal field development project. The Ladakh Lieutenant Governor R K Mathur hadstated it is a step towards sustainable development of Ladakh. It is an innovative method of achieving the goal of carbon–neutral Ladakh.

A few main points of the pact are:

- In Phase-1—1 megawatt (MW) power generation capacity shall be generated.
- Power will be supplied to the general public totally free of cost.
- In Phase-2—optimal number of wells shall be drilled.
- In Phase-3—The planned commercial project would be installed.
- The hot water from the spring could be used for indoor space-heating and also for hot water swimming pools, which would attract tourists.
- Making Ladakh a self-sustainable region.

References

1. https://simple.m.wikipedia.org/wiki/File:Historical_global_urban_-_rural_population_trends.png,File:Historicalglobalurban-ruralpopulationtrends.png
2. Renewable Energy Policy Market: Global Industry Analysis and Forecast (2023–2029) Trends, Statistics, Dynamics, Segmentation by Type, End-User, and Region. https://www.maximizemarketresearch.com/market-report/renewable-energy-policy-market/200155/
3. https://www.e-education.psu.edu/eme807/node/649
4. https://www.un.org/en/climatechange/raising-ambition/renewable-energy
5. EIA Predicts Global Energy Demand Growth [1.9.2]. https://economictimes.indiatimes.com/industry/energy/power/mnre-urges-states-to-provide-incentives-to-manufacturers/articleshow/75232158.cms
6. https://ourworldindata.org/energy-production-consumption
7. https://www.ren21.net/why-is-renewable-energy-important/
8. https://www.wec-italia.org/wpcontent/uploads/2016/10/worldenergyscenarios2016_fullreport.pdf
9. https://iea-etsap.org/E-TechDS/PDF/E15_Ren_integr_FINAL_Dec2013_GSOK.pdf
10. https://www.iea.org/reports/global-energy-review-2020/electricity
11. https://timesofindia.indiatimes.com/india/how-pandemic-is-changing-indias-energy-equations/articleshow/77931859.cms
12. Elavarasan RM, Shafiullah GM, Raju K, Mudgal V, Arif MT, Jamal T, Subramanian S, Sriraja Balaguru VS, Reddy KS, Subramaniam U (2020) COVID-19: Impact analysis and recommendations for power sector operation
13. https://www.statista.com/statistics/457821/europe-change-in-electricity-demand-covid-19/
14. https://simplystakeholders.com/renewable-energy-stakeholders/
15. https://irena.org/publications/2020/Jun/Post-COVID-Recovery
16. https://unfccc.int/process-and-meetings/the-paris-agreement

17. https://pib.gov.in/Pressreleaseshare.aspx?PRID=1525934
18. European Environment Agency (EEA) Report: Reuters, Brussels, updated on 18 Jan 2021 05:35 PM IST. Hindustan Times
19. https://www.eea.europa.eu/en/analysis/indicators/share-of-energy-consumption-from
20. https://mnre.gov.in/
21. https://www.exponent.com/knowledge/thought-leadership/2020/09/making-renewable-energy-usable-energy/?pageSize=NaN&pageNum=0&loadAllByPageSize=true
22. IPCC*(IPCC) in its synthesis report (AR5), Swedish Energy Agency. https://www.cowi.com/about/news-and-press/comparing-co2-emissions-from-different-energysources#:~:text=GHG%20emissions%20are%20measured%20in,emissions%20of%20different%20greenhouse%20gases.&text=Coal%3A%20740%2D1689%20g%20CO2e%2FkWh

Chapter 2
Waste-to-Energy (WtE)

2.1 Introduction

Waste is a natural by-product of the phenomena of life and growth of societies. It is viewed as unwanted or unusable material that has been disposed or discarded after primary use. Plants shed leaves, animals excrete. Humans in their day to day life create a boundless heap of waste of countless variety. The advancements in technology, changed standard of living has further increased the generation of waste not only in quantity but also types. Earlier we had more of food and agriculture waste but now the various types are E-Waste, Medi-Waste, Waste due to urbanisation as cardboard and plastic.

The high rise in fuel prices and depleting energy resources has compelled many countries to look for alternate sources of energy, for both economic and environmental reasons.

With growing public awareness about sanitation, and with increasing pressure on the government and urban local bodies to manage waste more efficiently, every country energy sector is poised to grow at a rapid pace in the years to come. The dual pressing needs of waste management and reliable renewable energy source are creating attractive opportunities for investors and project developers in the waste to energy sector.

The 2016 World Energy Council [1] reports state that according to the current rate of waste generation, global waste is estimated to reach 6 million tonnes/day by 2025. Therefore, increasing the Waste-To-Energy Technology adoption schemes would be the most effective method to reuse waste for various forms of energy generation and there by protect the environment.

N. Manohar et al., *Machine Learning Applications in Renewable Energy*, Green Energy and Technology, https://doi.org/10.1007/978-981-97-9939-8_2

2.2 Waste-to-Energy (WtE) Defined

According to the EPA (Environment Protection Agency) [2] the definition of Renewable Energy—"Renewable Energy is energy obtained from sources that are essentially inexhaustible, unlike natural gas, coal and oil, of which there is a finite supply."

According to the Department of Energy (DOE) [3]—"Renewable energy sources include: wood and other biomass, solar (Photovoltaic and Thermal), wind, geothermal, wastes [Municipal Solid Waste (MSW), Refuse-Derived Fuel (RDF), Landfill Gas (LFG)] and any other sources that are naturally or continually replenished." By definition, the DOE describes renewable energy as a "non-deplete-able source of energy."

Waste-to-Energy (WtE) is the process of generating energy in the form of electricity and/or heat from the primary treatment of waste, or the processing of waste into a fuel source. WtE is a form of energy recovery. Most WtE processes generate electricity and/or heat directly through combustion, or produce a combustible fuel commodity, such as methane, methanol, ethanol or synthetic fuels. Thus, Waste-to-Energy (WtE) is a process that converts waste to:

- Energy as heat, electricity
- Fuel as methane, gasoline
- Chemical products as alcohols, ammonia.

WtE refers to a family of technologies [4] that treat waste to recover energy in the form of heat, electricity or alternative fuels such as biogas. The scope of the term 'Waste-to-Energy' is very wide, encompassing a range of technologies of different scales and complexity. These can include the production of cooking gas in household digesters from organic waste, collection of methane gas from landfills, thermal treatment of waste in utility size incineration plants, co-processing of Refuse Derived Fuel (RDF) in cement plants or gasification.

2.3 Why Waste to Energy?

Most wastes that are generated find their way into land and water bodies without proper treatment, causing severe water [5] and air pollution. The problems caused by solid and liquid wastes can be significantly mitigated through the adoption of environment-friendly waste to energy technologies that will allow treatment and processing of wastes before their disposal and also generate various forms of energy as heat, gas and electricity.

The environmental benefits of waste to energy, as an alternative to disposing of waste in landfills, are clear and compelling. Waste to energy generates clean, reliable energy from a renewable fuel source, thus reducing dependence on fossil fuels, the combustion of which is a major contributor to GHG emissions.

These measures would reduce the quantity of wastes, generate a substantial quantity of energy from them, and greatly reduce pollution of water and air, thereby offering a number of social and economic benefits that cannot easily be quantified.

In addition to energy generation, waste-to-energy can fetch significant monetary benefits. Some of the strategic and financial benefits from waste-to-energy business are:

Profitability—If the right technology is employed with optimal processes and all components of waste are used to derive value, waste to energy could be a profitable business. When government incentives are factored in, the attractiveness of the business increases further.

Government Incentives—The government of India already provides significant incentives for waste to energy projects, in the form of capital subsidies and feed in tariffs. With concerns on climate change, waste management and sanitation on the increase (a result of this increasing concern is the newly formed ministry exclusively for Drinking Water and Sanitation), the government incentives for this sector is only set to increase in future.

Related Opportunities—Success in municipal solid waste management could lead to opportunities in other waste such as sewage waste, industrial waste and hazardous waste. Depending on the technology/route used for energy recovery, eco-friendly and "green" co-products such as charcoal, compost, nutrient rich digestate (a fertilizer) or bio-oil can be obtained. These co-product opportunities will enable the enterprise to expand into these related products, demand for which are increasing all the time.

Emerging Opportunities—With distributed waste management and waste to energy becoming important priorities, opportunities exist for companies to provide support services like turnkey solutions. In addition, waste to energy opportunities exist not just in India but all over the world. Thus, there could be significant international expansion possibilities for Indian companies, especially expansion into other Asian countries.

2.4 WtE Process

2.4.1 WtE Process Introduction

Waste is any material or product that is of no-more use to a person, house, Institution or any organization. The person, house, Institution or any organization that does not need the material is a Waste—**Dissipater** [6]. Thus, a waste-dissipater may be a single person or a group of persons belonging to a household, Institution or any Organization.

The waste can be processed to generate various form of energy such as heat, gas and electricity which can be used for the corresponding activities and thereby reduce

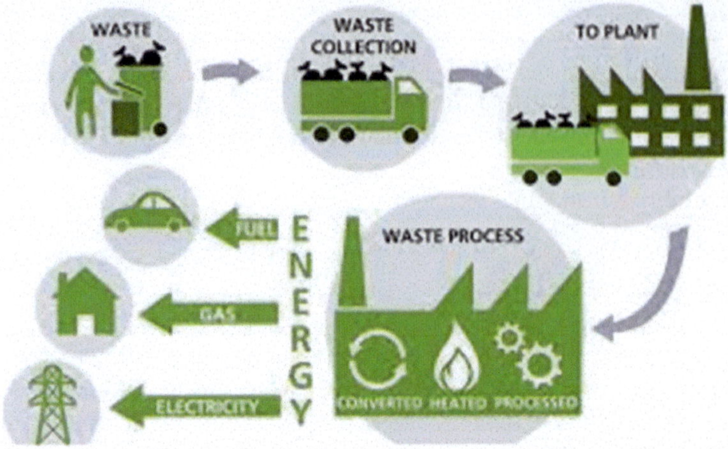

Fig. 2.1 Waste-to-energy system

the consumption of commercial energies. The wastes are of different types as solid, liquid and gases.

The stages of Waste-to-Energy process are:

1. Process Stage 1: Waste Collection
2. Process Stage 2: Waste Preparation for Conversion Process
3. Process Stage 3: Waste Conversion Process.

At every stage we need to consider the aspect of hygiene, pollution and health safety of all personnel. Figure 2.1 show to Waste-To-Energy System from the point of primary waste disposal to the final generation of energy.

Process Stage 1: Waste Collection

Waste is collected from the point of generation, in trucks and transported to the WtE Plant. At the entry point of the WtE plant the waste is weighed and unloaded onto the tipping floor.

Waste collection has three stages i. Primary Stage ii. Secondary Stage iii. Third Stage.

i. **Primary Stage**: It is at the dissipater location. The best practice for effective waste handling is segregation of waste [7] into wet and dry at the source, as in Fig. 2.2.

ii **Secondary Stage**: Secondary stage collection is done by the Utility Department such as the Municipal Corporations. The waste from the dissipater end is collected, and transported in trucks to the landfill. The waste that collects in a landfill causes forming landfill gas [8], which can be regenerated into sources of energy for the following energy needs:

Fig. 2.2 Segregation of waste into wet and dry

- Natural gas
- Fuel cells
- Vehicle fuel
- Steam
- Electricity.

iii **Third Stage**: At the landfill the waste is primarily processed [9], for size reduction and easy transport. The materials are segregated as paper, plastic, wood and bio-waste, Fig. 2.3. The segregated waste are then sent to the respective recycling plants. The segregated materials are compacted as in Fig. 2.4 for easy handling and transport.

The compacted waste is then transported to either recycling plants or waste-to-energy conversion plants.

Recent innovation is the scientists from NTU Singapore [10] have developed a technique to convert waste paper, from single-use packaging and bags, and cardboard boxes, into a crucial component of lithium-ion batteries for smart phones and electric vehicles.

Process Stage 2: Waste Preparation for Conversion Process

The input to the conversion process should be of proper size and type. The waste collected from the disposed site, has various materials of different shapes and sizes. The waste is inspected for unwanted material as nails, tin items which cannot be processed and separated. The waste that can be processed is then put into a tipping storage area where it is kept at negative air pressure.

Fig. 2.3 Segregation of waste

Fig. 2.4 Compacted waste

Size reduction is carried out to obtain a product, which is uniform and smaller in size, which can be housed in the combustion chamber. Only non- recyclable non-biodegradable high-calorific-value waste should be used as waste feed for WTE. As most of the waste sent to WTE plants is un-segregated, it has high inert content. Not suitable for burning in WTE plants. To burn it, additional fuel is required.

Process Stage 3: Waste Conversion Process

In the Stage 3 of the waste conversion process, the waste is subjected to one of the many techniques. **The various techniques are:**

i. Thermal Conversion as incineration
ii. Bio-Chemical Conversion as Anaerobic Digestion, Fermentation
iii. Thermo chemical Conversion as Pyrolysis, Gasification
iv. Electro-Chemical Conversion.

Figure 2.5 depicts Types of Inputs, Techniques and Output Energies. The Fig. 2.5 gives a clear understanding of the various types of feed stocks or input raw material to the WTE Process, the various types of WTE process or techniques and the forms of output energies.

During the process of the conversion, the various by-products which may be energy or fuel are:

i. Heat
ii. Steam
iii. Condensed water
iv. Gases as methane and Ashes.

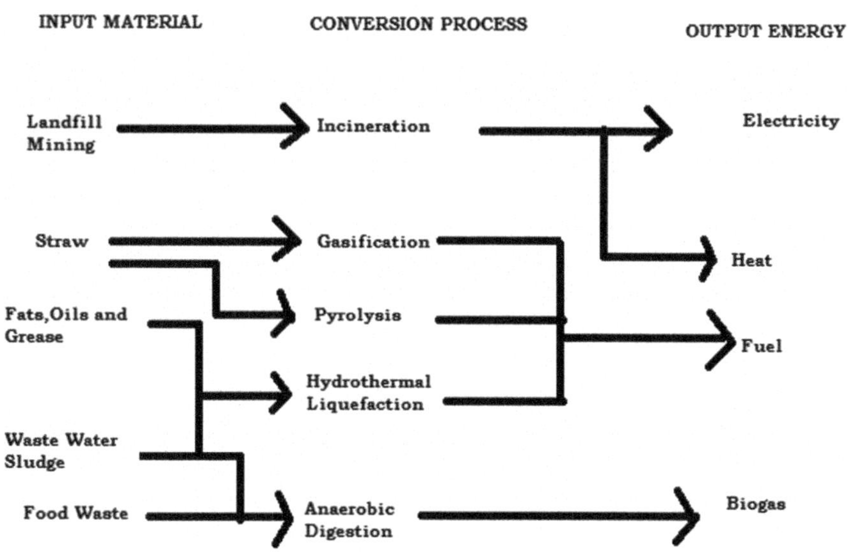

Fig. 2.5 Types of inputs, techniques and output energies

When the waste is fed into the combustion chamber, it is burnt at 2000 °C. The heat produced boils the water to generate steam. This steam is passed to the Turbine where in the rotor rotates to convert mechanical energy to electrical energy. One Ton of waste generates 550–750 KW of electricity.

2.5 Advantages of Waste-to-Energy Implementations

The advantages of waste to energy implementation [11] can be categorized into three as in Table 2.1.

2.6 Hurdles in Waste-to-Energy Implementations

Successful implementation of any new technology needs consideration of the hurdles in technical factor, socioeconomic and political aspect. The introduction of WtE technologies is often jeopardized by common obstacles such as:

- Missing tariff systems to fund investments and operation costs,
- Weak enforcement of environmental laws [12] and
- Limited qualified staff to run the installed systems in an efficient and effective manner.
- Lack of awareness in public.
- Lack of Social responsibility in public.

If such aspects are not taken into account, WtE projects risk failing at the cost of the municipality and local environment.

PESTEL analysis a strategic framework is most commonly used to evaluate the feasibility of a project. Traditionally, the framework was referred to as a PEST analysis [13], an acronym for **P**olitical, **E**conomic, **S**ocial, and **T**echnological. Presently, the framework is extended to include **E**nvironmental and **L**egal factors as well, which is very much essential for projects as Waste-To-Energy Plants. The various factors [14] under each category of PESTEL Analysis are presented in Table 2.2

The selection of the factors for assessment varies from country-to-country and region-to-region.

Table 2.1 Advantages of waste-to-energy implementation

Technical advantage	Economical advantage	Health/environmental advantage
They generate renewable energy	They lead to decentralized processing of waste	Conserves natural resources
Provides viable, effective alternative to fossil fuel	Generates revenue from recovery of recyclable materials	Creates a net negative GHG profile (methane, CO_2)
Uses readily available fuel source	Revenues from sale of recycled material and recovered energy (electricity, heat/steam/ash)	Reduces the need for new landfill space and emissions (lead, mercury, dioxins/furans, sulphuric and hydrochloric acids, volatile organic compounds, sulphur dioxide, nitrogen oxides, carbon monoxide, particulates)
Produces by products as heat and ashes	Creates local job opportunities	Filters out drug residues, heat-resistant bacteria, harmful chemicals
Supplies steam for process use	Helps to earn subsidies or avoid fines by meeting regulatory compliance targets	Improved waste management leads to a better quality of life for local populations, by improving hygiene conditions and reducing health risks related to illegal dumping and inadequate garbage collection
Supplies steam to chillers for air conditioning	Cross financing of MSW services through other local fees or taxes	They utilize waste effectively by generating energy through power or gas
Produces energy for desalination of seawater	Gate-fees when waste is delivered at a plant site	They provide eco-friendly solutions
Reduces transportation energy by treating waste locally rather than shipping to distant landfills	• Local or national subsidies • Revenues from national or international carbon funds (e.g. green climate fund) • Tax refunds and application of special feed-in tariffs for electricity produced from non-conventional sources such as waste material	Their emissions are in alliance with defined environment laws

Table 2.2 Factors under each category of PESTEL Analysis

Political factors	Economic factors	Social factors	Technological factors	Environmental factors	· Legal factors
• Government • Tax policies • Government regulation and deregulation • Special tariffs • Political action committees • Competition regulations • Government subsidies • Bilateral relationships • Import–export regulation/restrictions • Trade control	• Growth rate • Interest rate • Inflation rate • Availability of credit • Propensity of people to spend • Budget deficits • Gross domestic product trend • Unemployment trend • Stock market trends • Price fluctuations	• Acceptance by society • Social behaviour • Population size and growth rate • Immigration and emigration rates • Life expectancy rates • Age distribution • Wealth distribution • Social classes • Per capita income • Family size and structure • Lifestyles • Health consciousness • Average waste disposal • Attitude towards government • Attitude towards work • Buying habits • Ethical concerns • Cultural norms and values • Education level	• Availability of means and methods for new technology • Availability of skilled personnel • Training facilities • Technology incentives • Automation • R&D activity • Technological change • Access to new technology • Level of innovation • Technological awareness • Internet infrastructure • Communication infrastructure • Life cycle of technology	• Weather • Climate • Environmental policies • Climate change • Pressures from NGO's • Natural disasters • Air and water pollution • Recycling standards • Attitudes towards green products • Support for renewable energy	• Discrimination laws • Antitrust laws • Employment laws • Consumer protection laws • Copyright and patent laws • Health and safety laws • Education laws • Consumer protection laws • Data protection laws

2.7 Types of Wastes and Their Characteristics

2.7.1 Classification of Wastes

Growth of population, increasing urbanization and rising standards of living have contributed to an increase both in the quantity and variety of wastes generated by various activities. Classification of wastes is important to be able to manage waste. Wastes are classified based on:

i. **Source of generation** as Municipal Waste, Domestic Waste, Urban Waste, Rural Waste, Agricultural Waste, Industrial Waste, Medical Waste, E-Waste.
ii. **Physical State**: Solid Waste (SW),Liquid Waste (LW),Gas Waste (GW).
iii. **Level of Harm to Humans and Environment**: Hazardous and Non-Hazardous, Toxic and Non-Toxic, Bio-Degradable and Non-Bio-Degradable.

The classification categories are not exclusive; in that say domestic waste could be solid, liquid, or gas and toxic or non-toxic. We shall see the various examples of each category of waste and their sources.

Classification of waste is required for several factors as:

i. Determining the type, size and shape of the waste bin at the source of generation as house, school, shop etc.
ii. Determining the type, size, frequency of trips of the trucks to carry waste from generation source point to landfill, and landfill to WtE plant.
iii. To plan the environmental issues.
iv. Design of the WtE plant components as size reduction machines, technology for conversion from waste to energy or fuel.

In this text emphasis has been given on Solid waste as it is the most generated and technology is being developed to obtain various forms of energy as electricity, heat and biogas from it.

2.7.2 Major Classification of Solid Waste and Their Source

Solid waste is the material generated from various human activities and which is normally disposed off as useless and unwanted. The following Table 2.3 gives the various sources [15] of solid wastes and the type of wastes from each source.

Table 2.3 Sources of solid wastes and the type of wastes from each source

Source	Typical waste generators	Types of solid wastes
Residential	Single and multifamily dwellings	Food wastes, paper, cardboard, plastics, textiles, leather, yard wastes, wood, glass, metals, ashes, special wastes (e.g., bulky items, consumer electronics, white goods, batteries, oil, tires), and household hazardous wastes.), medi-waste
Industrial	Light and heavy manufacturing, fabrication, construction sites, power and chemical plants	Housekeeping wastes, packaging, food wastes, construction and demolition materials, hazardous wastes, ashes, special wastes based on the type of industry
Commercial	Stores, hotels, restaurants, markets, office buildings, etc	Paper, cardboard, plastics, wood, food wastes, glass, metals, special wastes, hazardous wastes
Institutional	Schools, hospitals, prisons, government centers	Same as commercial
Construction and demolition	New construction sites, road repair, renovation sites, demolition of buildings	Wood, steel, concrete, dirt, etc.
Municipal services	Street cleaning, landscaping, parks, beaches, other recreational areas, water and wastewater treatment plants	Street sweepings; landscape and tree trimmings; general wastes from parks, beaches, and other recreational areas; sludge
Process (manufacturing, etc.)	Heavy and light manufacturing, refineries, chemical plants, power plants, mineral extraction and processing	Industrial process wastes, scrap materials, off-specification products, slay, tailings
Agriculture	Crops, orchards, vineyards, dairies, feedlots, farms	Spoiled food wastes, agricultural wastes, hazardous wastes (e.g. pesticides)

2.8 Characteristics of Solid Waste

The characteristics of solid waste is required to be analyzed to:

- Select the waste-to-energy conversion technique,
- Identify the by-products that can be obtained,
- Analyze the final energy or fuel that can be obtained,
- Plan the process operating conditions,
- Plan for precautions to handle the by-products and residue,
- Study the environmental laws to be abided by.

The characteristics of solid waste are broadly classified [16] into physical and chemical characteristics. They are determined by conducting suitable tests in various laboratories as per standards.

2.8.1 Physical Characteristics of Solid Waste

Information and data on the physical characteristics of solid wastes are important for the selection and operation of equipment and for the analysis and design of disposal facilities. The major physical characteristics of solid waste are:

(1) Density
(2) Size distribution of components
(3) Moisture content
(4) Calorific value
(5) Field Capacity
(6) Permeability
(7) Compressibility

Other characteristics which may be used in making decision about solid waste management are:

(1) Colour
(2) Voids
(3) Shape of components
(4) Optical property
(5) Magnetic properties and
(6) Electric properties

The following physical characteristics are to be studied in detail.

1. **Density**

Density of waste, i.e., its mass per unit volume (kg/m^3), is a critical factor in the design of a solid waste management system. The parameter is the basis for the design of sanitary landfills, storage, types of collection and transport vehicles, etc.

Efficient operation of a landfill demands compaction of wastes to optimum density. Any normal compaction equipment can achieve reduction in volume of wastes by 75%, which increases an initial density of 100–400 kg/m^3. In other words, a waste collection vehicle can haul four times the weight of waste in its compacted state than when it is uncompacted. Significant changes in density occur spontaneously as the waste moves from source to disposal, due to scavenging, handling, wetting and drying by the weather, vibration in the collection vehicle and decomposition.

2. **Size**: The size and size distribution of the component materials in solid wastes are an important consideration for:

i. Recovery of materials especially with mechanical means such as trommel screens and magnetic separators.
ii. Significance in the design of mechanical separators and shredders.

Size distribution is expressed as a plot of particle size (mm) against percentage, less than a given value. The size of waste component may be defined by one or more of the following measures based on the shape:

$$S_c = l \tag{2.1}$$

$$S_c = (l + w)/2 \tag{2.2}$$

$$S_c = (l + w + h)/3 \tag{2.3}$$

$$S_c = (l\,x\,w)^{1/2} \tag{2.4}$$

$$S_c = (l\,x\,w\,x\,h)^{1/3} \tag{2.5}$$

where S_c is the size of the component in mm
L is the Length in mm
L is the Length in mm
W is the Width in mm
H is the Height in mm

3. **Moisture Content**

Moisture Content is the ratio of weight of water to the total wet weight of waste. It can be calculated:

$$\%\,\text{Moisture Content} = \frac{\text{Wet Waste Weight} - \text{Dry Waste Weight}}{\text{Wet Waste Weight}} \times 100 \tag{2.6}$$

Moisture increases the weight of solid wastes, and thereby, the cost of collection and transport. Moisture content is a critical determinant in the feasibility of waste

Table 2.4 Heating value of municipal solid waste components

S. No.	Component	Heating values	
		Range	Typical
1	Food waste	3500–7000	4500
2	Paper	11,500–18,500	16,500
3	Cardboard	14,000–17,500	16,000
4	Plastic	28,000–37,000	32,500
	Textiles	**15,000–20,000**	**17,500**
	Rubber	**21,000–28,000**	**18,500**
	Leather	**15,000–20,000**	**17,500**
	Garden trimmings	**2300–18,500**	**6500**
	Wood	**17,500–20,000**	**18,500**
	Glass	**120–240**	**140**
	Non ferrous metal	**240–1200**	**700**
	Ferrous metal	**240–1200**	**700**
	Dirt, ash and bricks	**2300–11,500**	**7000**
	Municipal solid waste	**9500–13,000**	**10,500**

treatment by incineration, because wet waste consumes energy for evaporation of water and in raising the temperature of water vapour. This increases the process time. Wastes should therefore be shielded from getting wet due to rainfall or other activities that cause falling of water onto waste.

4. **Calorific Value**

Calorific value is the amount of heat generated from combustion of unit weight of solid waste, expressed as Kcal/Kg. The calorific value can be determined using Bomb Calorimeter. Table 2.4 gives the heating value of various municipal solid waste components—range and typical values.

5. **Field Capacity**

The field capacity of solid waste is the amount of moisture that can be retained in a waste sample subject to the downward pull of gravity. The field capacity of waste materials is important in determining the formation of leachate in landfills. The field capacity varies with the degree of applied pressure and the state of decomposition of the waste. The field capacity of uncompacted commingled wastes from residential and commercial sources is in the range of 50–60%.

6. **Permeability**

- Permeability is defined as the hydraulic conductivity of compacted waste. It is an important physical property as it governs movement of liquid and gases in landfill. It depends on pore size,

- Surface area and
- Pore size distribution.

Permeability is inversely related to density, implying that denser refuse is less permeable. The reported range of permeability of refuse is 10–1 to 10–5 cm/s.

7. **Compressibility**

Compressibility is the degree of physical changes in solid waste when subjected to pressure.

2.8.2 *Chemical Characteristics of Solid Waste*

Chemical composition of solid wastes is important to evaluate alternative processing, recovery options and determining the quantity of fuel which be be the final product of a process or by-product. Typically waste is considered as combination of combustible and non-combustible components. If solid waste is to be used as a fuel or for any other use we should know its chemical components. The various chemical components of solid waste and their source is tabled in Table 2.5

The test to be conducted to determine the chemical characteristics of solid waste are: i. Proximity Analysis ii. Ultimate Analysis

Proximate Analysis

Proximate analysis is conducted to determine the moisture, ash, volatile matter, and fixed carbon contents of solid wastes.

(a) **Moisture**: The moisture content of the sample is determined by drying 1 g of sample at 1050 °C for one hour. Weight loss is expressed in % of initial weight of sample as

Table 2.5 Chemical components of solid waste and their source

Chemical component of solid waste	Source of the component
Lipids	Fats, oils and grease
Carbohydrates	Food sources rich in starch and celluloses
Proteins	Compounds containing carbon, hydrogen, nitrogen and oxygen and organic acid with amino groups. They are primarily found in food and garden wastes
Natural fibers	Products contain cellulose and lignins that are relatively resistant to biodegradation. These are found in paper products, food and yard wastes
Synthetic organic materials	Plastics
Non-combustibles	Glass, ceramics, metals, dust and ashes

Table 2.6 Proximate analysis of Municipal Solid Waste

S. No.	Components	Value percent	
		Range	Typical
1	Moisture	15–40	20
2	Volatile matter	40–60	53
3	Fixed carbon	5–12	7
	Glass, ash and metal	15–30	20

$$\text{Moisture}\ (\%) = (\text{weight loss}/\text{weight of sample}) \times 100 \quad (2.7)$$

(b) **Volatile matter**: Volatile matter is the weight loss obtained on heating 1 gm sample of substance at 950 °C for 7 min in the absence of air

$$\text{Volatile Matter}(\%) = (\text{weight loss due to VM}/\text{weight of sample}) \times 100. \quad (2.8)$$

(c) **Ash**: Ash is the weight of the residue obtained after complete combustion of one gram of the substance at 700–750 °C.

$$\text{Ash}(\%) = (\textit{weight}\ \text{of residue}/\text{weight of sample}) \times 100 \quad (2.9)$$

(d) **Fixed carbon**: Fixed carbon is the material, other than ash, that does not vaporize when heated in the absence of air. It is usually determined by subtracting the sum of the first three values (i.e.) moisture, ash, and volatile matter in weight percent from 100%.

$$\text{Fixed carbon}(\text{W\% wet basis}) = 100 - (\%\text{M} + \%\text{Ash} + \%\text{VM}) \quad (2.10)$$

Table 2.6 summarizes the proximate analysis of typical municipal solid waste generated in India.

Ultimate Analysis

The ultimate analysis of solid waste is conducted to determine:

- The percent of C (carbon), H (hydrogen), O (oxygen), N (nitrogen), S (sulphur), and ash.
- The results of the ultimate analysis are used to characterize the chemical composition of the organic matter in MSW.
- They are also used to define the proper mix of waste materials to achieve suitable C/N ratios for biological conversion processes.

2.9 Characteristics of Liquid Waste

Liquid waste or waste water is any water that has been used and cannot be reused for any other purpose. However, waste water when treated in a sewerage treatment plant (STP) and processed, can be used for purposes as watering of plants. Thus the processed water of STP, serves two purposes for trees, plants and agricultural fields. The two purposes are serving as manure and liquid food for greenery.

The various characteristics of liquid waste is as in Fig. 2.6.

2.9.1 Physical Characteristics of Liquid Waste

- **Turbidty** in wastewater is the **level of cloudiness or the presence of suspended particles in the wastewater sample. It is usually measured in FTU-Formazin Turbidity Unit.**
- **Color and Odor:**

The **color of fresh water** Fig. 2.7 varies with the ambient conditions in which that water is present. While relatively small quantities of water appear to be colorless, pure water has a slight blue color that becomes deeper as the thickness of the observed sample increases.

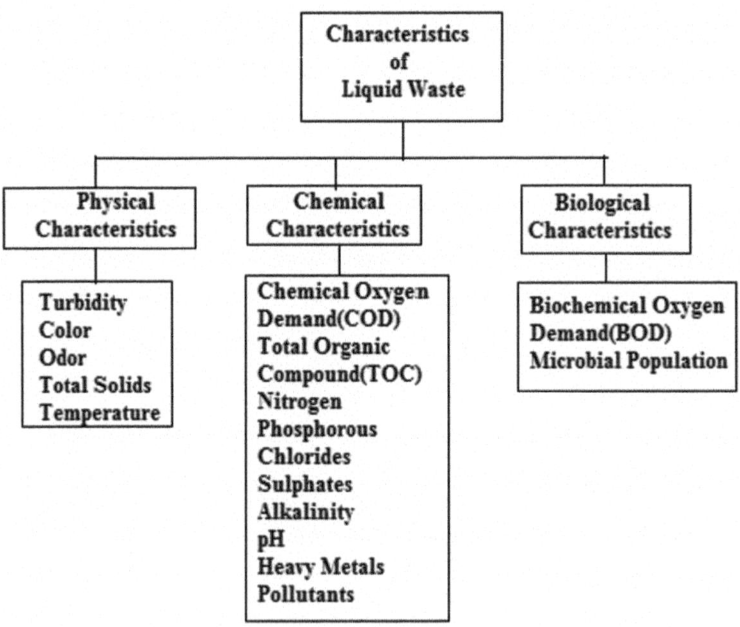

Fig. 2.6 Classification of characteristics of liquid waste

Fresh water in small quantities as in glass appears colorless | Fresh water in large quantities appears blue | Waste Water

Fig. 2.7 Color of fresh and waste water

The color of wastewater Fig. 2.7 containing dissolved oxygen (DO) is normally gray. Black-colored wastewater usually accompanied by foul odors and containing little or no DO is said to be septic. Fresh sewage is light brownish grey colour. Sewage water over time is converted to dark grey and black color due to anaerobic activities, known as stale or septic color. The grey, dark grey and black color is due to formation of sulfide produced under anaerobic conditions reacts with the metals present in wastewater.

- **Total solids**: Solids comprise matter suspended or dissolved in water and wastewater.Solids are divided into several different fractions and their concentration provide useful information for characterization of wastewater, selection and control of treatment processes.

Total solids is the sum of total suspended solids and total dissolved solids. Each of these groups can further be divided into volatile and fixed fractions. Total solids can be expressed in mg/L.

- **Temperature**: The normal temperature of sewage water is commonly higher than fresh water supply due to domestic and industrial activities. Depending on geographical location, the mean annual temperature of sewage water is in the range of 10–21 °C. Temperature of sewage water is an important parameter as it effects the chemical reaction rates, aquatic life and environment. Increase temperature can cause a change in fish species that are present in water bodies.

2.9.2 Chemical Characteristics of Liquid Waste

- **COD (Chemical Oxygen Demand): Chemical Oxygen Demand (COD) is a test that measures the amount of oxygen required to chemically oxidize the organic material and inorganic nutrients, such as Ammonia or Nitrate present in water O.M + O_2 + Microorganisms.**

- **Total Organic Carbon (TOC) is a measure of the total amount of carbon in organic compounds in pure water and aqueous systems.**

2.9.3 Biological Characteristics of Liquid Waste

- **Biochemical oxygen demand (BOD)** is an analytical parameter representing the amount of dissolved oxygen consumed by aerobic bacteria growing on the organic material present in a water sample at a specific temperature over a specific time period.

2.10 Waste Disposal Management

2.10.1 Introduction

Generation of waste is inherent of any individual or organization activity. However, proper disposal of the waste and handling requires planning and management. Every individual, Management of organizations and the Government need to plan at their own level. Organization could be an industry, educational institution, commercial establishment or any establishment run by a group of people.

Government in all countries has its own waste management policies and procedures. The management of organizations shall then prepare their own plans of waste management in line with that of the government.

The process of waste management involves treating of solid and liquid wastes. Since, the two types are of different physical states, their handling methods should be different. Figure 2.8 depicts the Waste Disposal Management hierarchy.

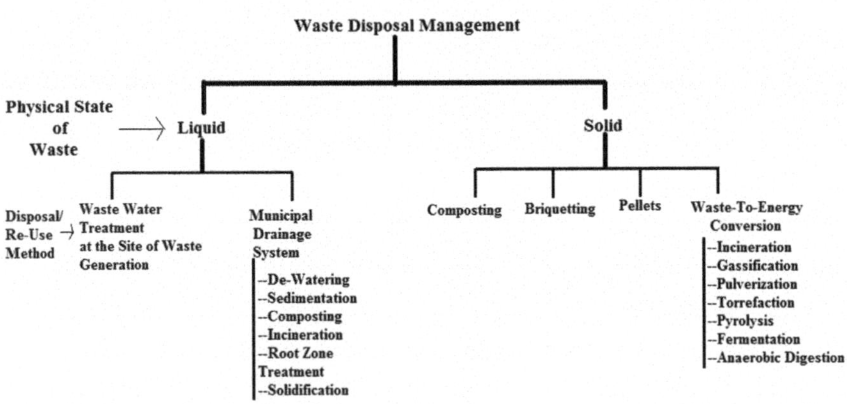

Fig. 2.8 Waste disposal management hierarchy

Based on the physical state, wastes are of two types which can be handled—solid and liquid. A third state waste which is gaseous does exist but it is difficult to handle. With advancement in technology, liquid waste can be now reprocessed, and the treated waste can be used for watering of plants, washing of common areas a portico etc. The liquid sewerage waste can be be processed through—de-watering, sedimentation, composting, incineration root zone treatment, solidification.

Solid waste can go through the process of—composting, briquetting, pelleting and one of several waste-to-energy conversion techniques. The output of all the processes that the liquid and solid waste are processed through is either fuel or energy.

2.10.2 Liquid Waste Disposal

Liquid waste can be disposed or converted into useful form by two process channels:

i. Waste water treatment at the site of waste generation
ii. Municipal drainage system.

2.10.2.1 Waste Water Treatment

Waste Water treatment [17, 18] comes under the Principle of Re-Cycle and Re-Use of the 3R and 4R Principle of Waste Management, that proposes zero-waste hierarch, Fig. 2.9. Waste water is generated from human activities and natural occurrence as rain. It is also called as sewage. The sewage from rain drains away into water bodies as rivers, ponds etc. However, waste water from human activities which contain pollutants can be contained and treated, purified to a certain extent for reuse. This type of sewage is generally categorized by the source point of generation as: domestic sewage, industrial sewage and storm sewage.

The waste water treatment process is presented in the form of flow chart [19] in Fig. 2.10. Basically, the process has three stages: pre-treatment stage, primary stage and secondary stage. The flow-chart though self explanatory, the various activities are explained below:

Filtering: The waste water is initially, as a pre-treatment process, is sent through the filter unit. The filter unit consists of screens as bar screen and mesh screen to remove the large objects and non-degradable objects.

Sand Collector: In the sand collector, any suspended sand particles are allowed to settle down by reducing the velocity of water. This is a physical process.

Bio-Chemical Treatment: Biological or bio-chemical wastewater treatment often is used as a secondary treatment process to remove material remaining after primary treatment with processes including dissolved air flotation (DAF).

Biological wastewater treatment uses natural processes for the decomposition of organic substances. Biological treatments rely on bacteria, nematodes, or other small organisms to break down organic wastes using normal cellular processes. Wastewater

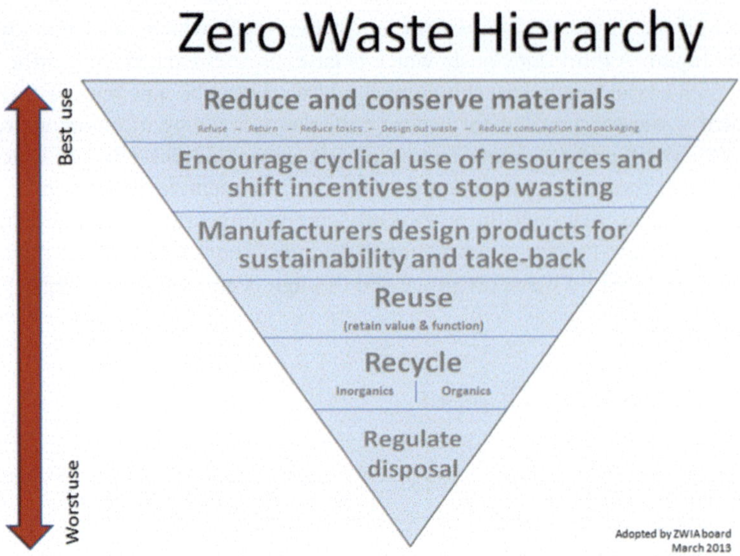

Fig. 2.9. 3R, 4R principle and zero-waste hierarchy

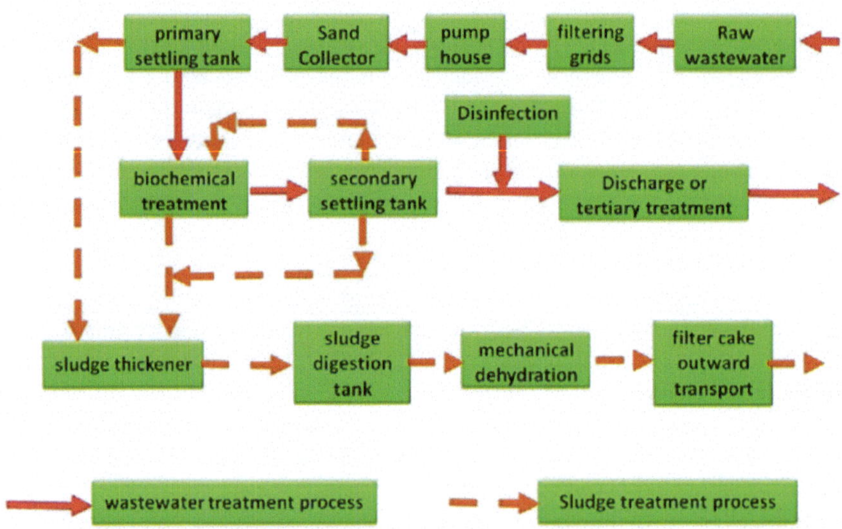

Fig. 2.10 Waste water treatment process

typically contains a mixture of organic matter, such as garbage, wastes, and partially digested foods. It also may contain pathogenic organisms, heavy metals, and toxins.

The goal of biological wastewater treatment is to create a system in which the results of decomposition are easily collected for proper disposal.

Biological treatment usually is divided into aerobic and anaerobic processes. "Aerobic" refers to a process in which oxygen is present, while "anaerobic" describes a biological process in which oxygen is absent.

MABR Treatment

Advancements in technology have transformed boil-chemical processes of waste water treatment. One such process is the membrane aerated biofilm reactor [20] (MABR), which uses 90% less energy for aeration. In just a few years, MABR has developed into a mature technology.

Disinfection: Water disinfection is the removal, deactivation or killing of pathogenic microorganisms. Microorganisms are destroyed or deactivated, resulting in termination of growth and reproduction. Diseases such as typhoid and paratyphoid fevers, cholera, salmonellosis, and shigellosis can be prevented by using treatments that significantly lower the overall amount of live bacteria in the water.

Sludge Disposal: This is the final stage of waste water treatment process. Sludge is composed of both organic and inorganic materials, a large concentration of plant nutrients, organic chemicals, as well as pathogens. Therefore, it is extremely important to properly treat such sludge in order to minimize its environmental repercussions. Steps in sludge treatment and disposal are:

Step 1—Sludge Thickening

In this step, the sewage sludge is thickened in a gravity thickener to reduce its overall volume, thus enabling the easy handling of the sludge.

Step 2—Sludge Digestion

This is a biological process in which the organic solids present in the sludge are decomposed into stable substances. This process also helps reduce the total mass of solids, while destroying any present pathogens to enable easy dewatering.

Step 3—Dewatering

After retrieving useful gases and other byproducts, the remaining sludge is then dewatered before final disposal.

Step 4—Disposal

Once the sludge has been effectively dewatered, it can be buried underground in a sanitary landfill or can be used as a fertilizer, depending on its chemical composition. In cases where the sludge is too toxic to be reused or buried, it can be incinerated and convert into ash.

2.10.2.2 Municipal Drainage System

De-Watering: Sludge de-watering is minimizing waste by volume to prepare for effective disposal.

Sedimentation: Sedimentation [21], Fig. 2.11 is a physical water treatment process using gravity to remove suspended solids from water.

Composting: Composting of waste water residual is a bio-thermal aerobic process that decomposes the organic portion of the residuals. The composting process reduces the organic material in the residual by approximately 25%. During composting the

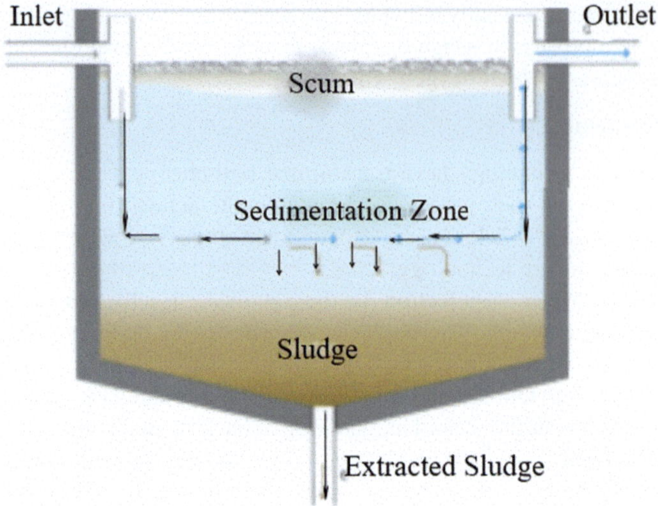

Fig. 2.11 Sedimentation

heat generated by the decomposition of the organic portion of the residuals reduces the moisture content of the residual, stabilizes it and renders the residual harmless by transforming it into a usable bio-solid.

Incineration: Incineration of sludge waste, derived from waste water, uses thermal treatment to oxidize organic matter contained in the sludge. Incineration is carried out in a furnace using oxygen and high temperatures (over 850 °C) resulting in total, or near-total, oxidisation of the organic material. It is a controlled combustion process with the primary objective of volume reduction and energy recovery from the waste stream.

Root Zone Treatment: The root zone/filter plant is a biological filter [22], where the biological treatment of waste water takes place in a soil volume, which is penetrated by roots. The root network is composed of suitable plant species. It is more beneficial if the optimum varieties of these species are selected. Figure 2.12 shows one such root zone waste water treatment plant.

Solidification: Solidification is the processes of making a waste to form a solid material and to restrict the migration of contaminant by decreasing the available surface area, which is exposed to leaching, when contacting liquids and also by coating the waste with low-permeability materials. Solidification involves mixing a waste with a binding agent, which is a substance that makes loose materials stick together. Common binding agents include cement, asphalt, fly ash, and clay. Water must be added to most mixtures for binding to occur; then the mixture is allowed to dry and harden to form a solid block.

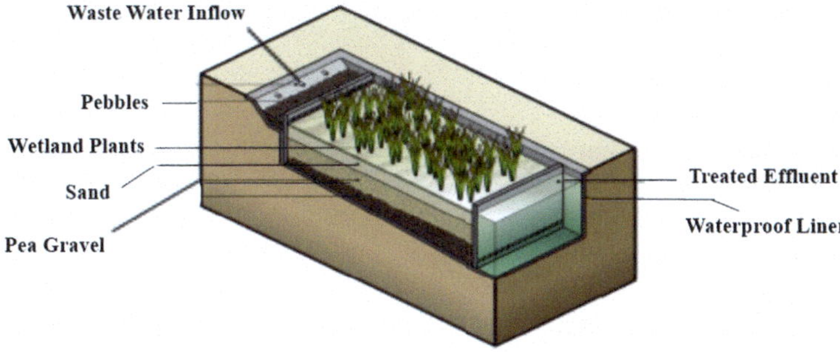

Fig. 2.12 Root zone waste water treatment plant

2.10.3 Solid Waste Disposal

As seen in Fig. 2.8, the solid waste disposal process channels are:

i. Composting
ii. Briquetting
iii. Pellets
iv. Waste-To-Energy.

Composting: Composting is a process by which organic wastes are broken down by microorganisms, generally bacteria and fungi, into simpler forms. The microorganisms use the carbon in the waste as an energy source. Heat generated during the process kills many unwanted organisms such as weed seeds and pathogens. Advantages of composting include reduction of waste volume, elimination of heat-killed pests, and the generation of a beneficial and marketable material. The end product of composting, is a good fertilizer for plants and agricultural crops.

Briquetting: Briquetting [23] process is the conversion of agricultural waste into uniformly shaped briquettes Figure 2.13, that are easy to use, transport, and store. Briquetting involves reducing moisture content in the organic waste, which is shredded and the biomass is compressed at high temperature using a binding agent. Figure 2.13 is the flow chart of briquetting process.

Pellets: Pelletization of municipal solid waste is the processes of segregating, crushing, mixing high and low heat value combustible waste material and solidifying it to produce fuel pellets Fig. 2.14, also known as Refuse Derived Fuel (RDF).

The calorific value of raw municipal solid waste is around 1000 kcal/kg while that of fuel pellets is 4000 kcal/kg. On an average, about 15–20 tons of fuel pellets can be produced after treatment of 100 tons of raw garbage. Since pelletization enriches the organic content of the waste through removal of inorganic materials and moisture, it is a very effective method for preparing an enriched fuel feed for other thermo-chemical processes like pyrolysis, gasification, and incineration.

Waste-to-Energy: The various waste-to-energy techniques are discussed in detail in the next Chapter.

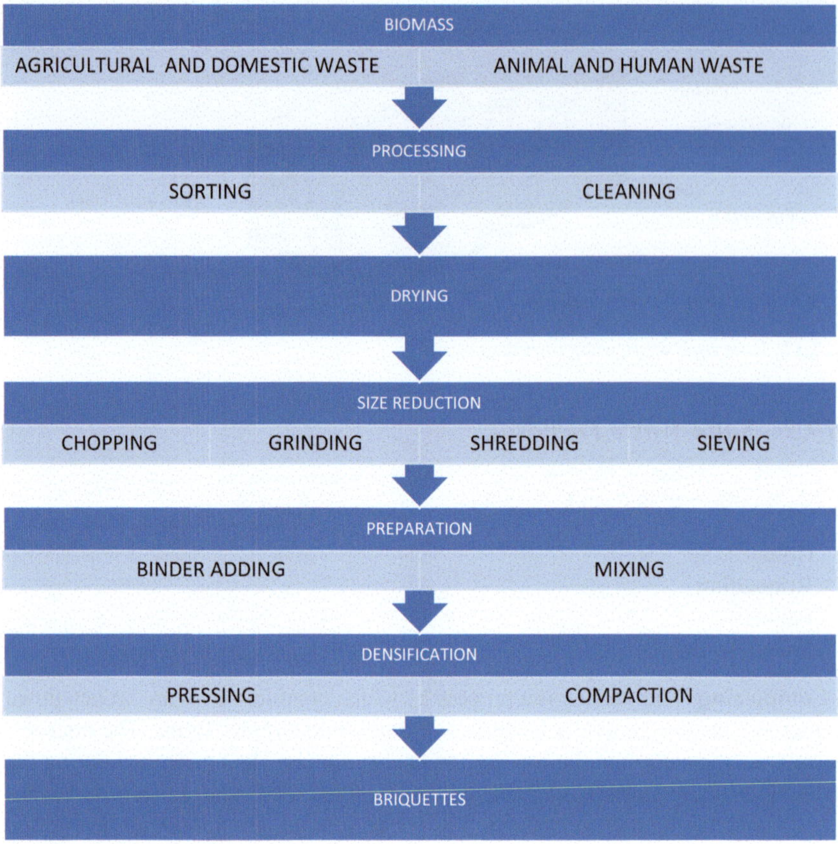

Fig. 2.13 Flow chart of briquetting process

Fig. 2.14 Biomass briquettes and biomass pellets

2.11 Methods of Converting Waste to Energy

Waste to Energy technologies convert waste into more useful forms of energy or fuel. The energy may be the end product of the process or a by-product. Most technologies provide at least one form of energy and a fuel. These wastes to energy processes include thermal conversions such as combustion (incineration), pyrolysis, gasification, or biological treatments such as anaerobic digestion and fermentation, etc., and various combinations of the above. All these techniques depend on:

- Quality of waste,
- The quantity of waste,
- Various different components in the waste,
- Aspired form of energy,
- Customer requirements,
- Rules,
- Standards of the environmental ministry of the country,
- Financial conditions of the private or government companies.
- Social conditions.

With increasing global consumerism and population growth leading to an increase in the levels of waste materials we produce, waste management technologies are now helping to find alternatives to landfills by converting as much of the mixed general waste back to be recycled or reused as possible in an attempt to turn waste into energy and waste into fuels.

Sending residual wastes direct to landfill, would increase pollution and also stages of processing; instead if advanced conversion technologies coupled with advanced pollution control systems are employed the waste which has high calorific content, can give us several fuels, energy and thereby also clean environment. Advanced waste-to-energy technologies can be used to produce biogas (methane and carbon dioxide), syngas (hydrogen and carbon monoxide), liquid bio-fuels (ethanol and biodiesel), or pure hydrogen. Just as oil, coal and gas are used as fuels in a fossil fuel fired power stations, these alternative bio-fuels can also be converted into electricity.

Converting **Waste to Energy** has many advantages too by not only reducing the amount of landfill dumping, but by reducing the amount of greenhouse-gas emissions and pollution we pump into the atmosphere each and every year as well as reducing our dependence on non-renewable fossil fuels. However, many wastes to energy conversion technologies are designed to handle only one or a few types of waste and can be difficult to fully separate different types of waste materials.

Advances in non-incineration conversion technologies and methods like pyrolysis and thermal gasification are providing ways of generating clean energy from waste materials that avoid many of the pollution concerns around conventional incineration and combustion. Today, we have the technologies and options available to us to separate the bio-waste which should be recycled, from the waste that can be used as a valuable and future energy source turning waste and other renewable waste fuels into clean energy.

2.11.1 Waste-to-Energy Techniques Classification

There are various technologies [1] to process waste. Under each technology there are several processes. The various technologies, the processes under each technique, and the output from each process are indicated in Fig. 2.15. Figure 2.15 depicts the taxonomy of various waste-to-energy conversion techniques.

The WtE Techniques are broadly classified into two categories:

1. Thermal Technologies
2. Non-Thermal Technologies.

The thermal technology encompasses:

i. Thermal Combustion
ii. Mechanical and Thermal
iii. Thermo-Chemical.

Non-thermal technology is bio-chemical.

Thermal Combustion—Combustion refers to the rapid oxidation of fuel accompanied by the production of heat, or heat and light. Complete combustion of a fuel is possible only in the presence of an adequate supply of oxygen. This is the oldest, most common thermal process using a wide variety of waste. Municipal and household waste is directly combusted in large waste to energy incinerators as a fuel with minimal processing known as mass burning. The earliest waste combustion systems were simple incinerators which produced heat and carbon dioxide, along with a variety of other pollutants but had no energy recovery facilities. The present modern combustion systems as shown in Fig. 2.16, the heat energy generated from the process is used to turn water into steam in a boiler, which is then used to power steam-turbine generators to produce electricity.

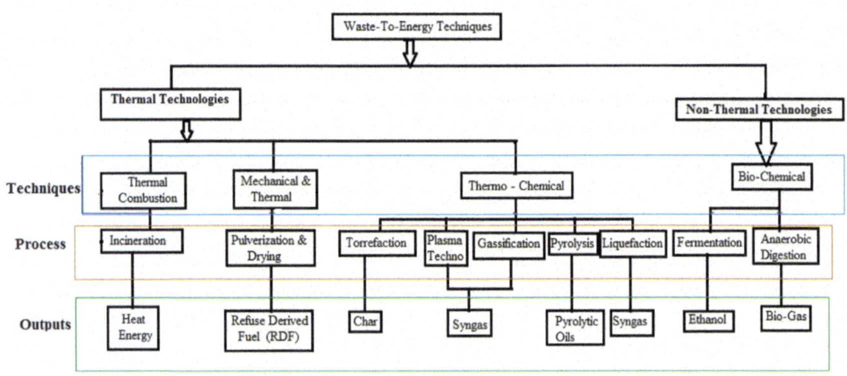

Fig. 2.15 Techniques, processes and outputs

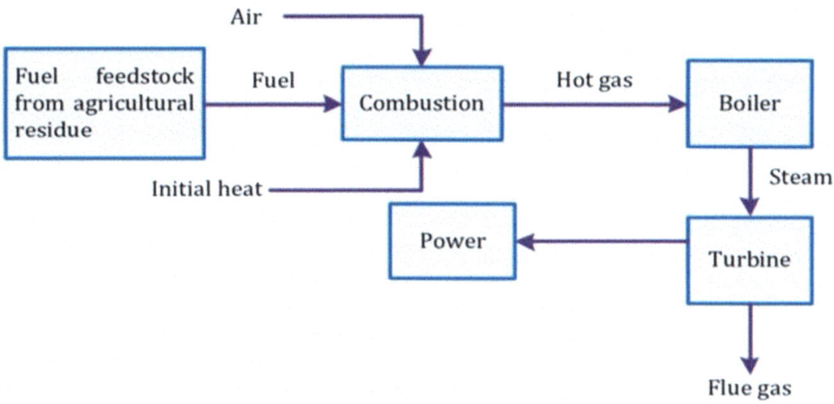

Fig. 2.16 Electricity from combustion

Thermal combustion process is incineration. Incineration is the high-temperature burning or rapid oxidation of a waste most modern waste incinerators now incorporate heat recovery systems and air-pollution control systems. They are efficient, have low environmental footprint and deliver quality heating of the product. The measures of combustion efficiency are: available heat, process efficiency and overall efficiency.

Incineration

Incineration is a waste treatment process that involves the combustion of substances contained in waste materials. Industrial plants for waste incineration are commonly referred to as **waste-to-energy** facilities. Incineration of waste materials converts the waste into ash, flue gas and heat. The ash is mostly formed by the inorganic constituents of the waste and may take the form of solid lumps or particulates carried by the flue gas.

Mechanical and Thermal Technique

Thermal technologies:

- Gasification: produces combustible gas, hydrogen, synthetic fuels
- Thermal depolymerization: produces synthetic crude oil, which can be further refined
- Pyrolysis: produces combustible tar/biooil and chars
- Plasma arc gasification or plasma gasification process (PGP): produces rich syngas including hydrogen and carbon monoxide usable for fuel cells or generating electricity to drive the plasma arch, usable vitrified silicate and metal ingots, salt and sulphur.

Non-thermal technologies:

- Anaerobic digestion: Biogas rich in methane
- Fermentation production: examples are ethanol, lactic acid, hydrogen

- Mechanical biological treatment (MBT)

  MBT + Anaerobic digestion
  MBT to Refuse derived fuel.

Thermo Chemical Techniques

Torrefaction

Torrefaction is a thermo-chemical process used to produce high-grade solid bio-fuels from various streams of woody biomass or agro residues. The end product is a stable, homogeneous, high quality solid bio-fuel with far greater energy density and calorific value than the original feedstock.

Biomass torrefaction involves heating the biomass to temperatures between 250 and 320 °C in a low-oxygen atmosphere. When biomass is heated at such temperatures, the moisture evaporates and various low-calorific components (volatiles) contained in the biomass are driven out. During this process mainly the hemi-cellulose in the biomass decomposes which transforms the biomass from a fibrous low quality fuel into a product with excellent fuel characteristics.

The advantages of torrefied biomass over untreated biomass are:

- Has higher calorific value
- They are more homogeneous
- Have higher bulk density
- Excellent grindability
- Durability
- Hydrophobic nature/water resistance
- No biological activity.

Plasma Technology: Plasma arc gasification (PAG), waste-treatment technology uses a combination of electricity and high temperatures to turn municipal waste (garbage or trash) into usable by-products without combustion (burning). The by-products of plasma arc gasification are syngas, slag and residual heat.

Gasification—Gasification is a thermo chemical process that converts solid wastes into a mixture of combustible gases. Steam or the oxygen in the air is reacted at high temperature with the available carbon in the waste material to produce gases such as carbon monoxide, hydrogen and methane. The gasification process produces a syngas (hydrogen and carbon monoxide) which is used for generating electricity power.

Incineration is the burning of waste fuels in an oxygen rich environment, where as gasification is the conversion of waste materials that takes place in the presence of limited amounts of oxygen. Incineration of waste to energy converts the fuel waste into energy directly on-site, thermal gasification of the waste materials allow the production of a gaseous fuel that can be easily collected and transported.

Pyrolysis—Pyrolysis is the thermal degradation of organic waste in the absence of free oxygen to produce combustible gases. Pyrolysis uses heat to break down organic materials in the absence of oxygen. Materials suitable for pyrolysis processing include coal, animal and human waste, food scraps, paper, cardboard, plastics and rubber. The pyrolytic process produces oil which can be used as a synthetic bio-diesel fuel or refined to produce other useful products.

Although pyrolysis technology has been around for a long time, its application to biomass and waste materials is a relatively recent development as pyrolytic products are more refined and therefore can be used with greater efficiency. A common byproduct of pyrolysis is a kind of fine-grained bio-charcoal called "bio-char", which retains most of the carbon and nutrients contained in biomass so can be used as a soil enhancement to increase soil productivity. In Fig. 2.17 we can see the output products of pyrolysis are oils, gas, char and some metals. The figure also depicts come applications of the output products. From this it can be understood that by processing waste through pyrolysis we can get some very useful products.

Liquefaction

Liquefaction is the thermo chemical conversion process, that converts biomass into bio-oil, bio-char and gases as syngas. The process is usually carried out in water or another suitable solvent at 25–400 °C under pressures of 5–25 MPa.

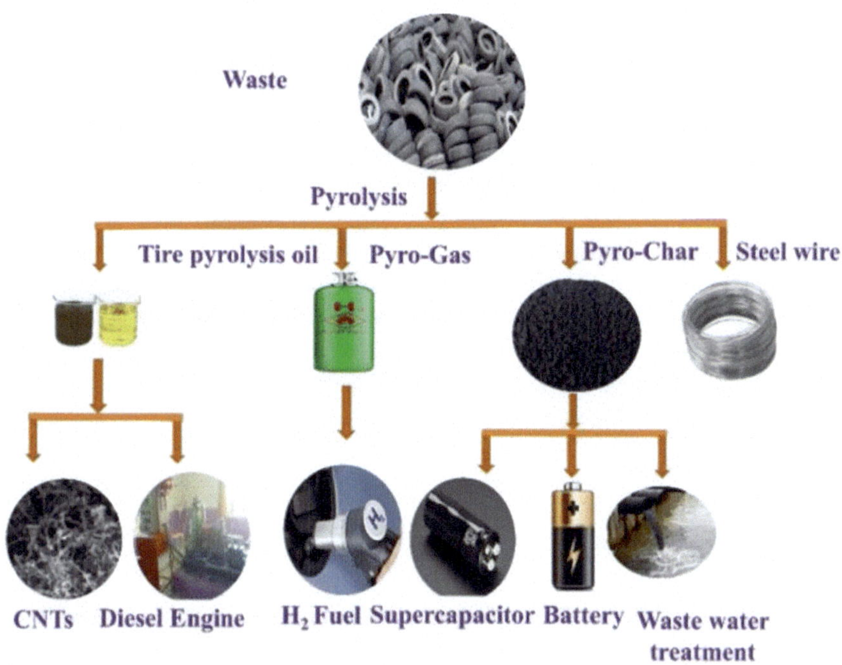

Fig. 2.17 Pyrolysis output products and their application

Bio-Chemical Techniques

Fermentation

Fermentation is the process operated by microorganisms to break down organic compounds to obtain energy through anaerobic metabolism. The fermentation process allows conversion of waste materials containing lipids, proteins and carbohydrates. Organic substrates serve both as electron donors and acceptors. The major fermentation output and by-products are volatile fatty acids such as acetic, propionic and butyric acid, and alcohols such as ethanol and butanol.

Anaerobic Digestion—Land filling is still the primary method of disposal of municipal solid waste and if left undisturbed, landfill waste produces significant amounts of gaseous byproducts, consisting mainly of:

- Carbon dioxide (CO_2) and
- Methane (natural gas, CH_4).

These landfill gases or "biogas" is produced by the anaerobic (oxygen-free) digestion of organic matter.

Anaerobic digestion to produce biogas can either occur naturally producing a landfill gas, or inside a controlled environment like a biogas digester. A digester is a warmed, sealed, airless container where bacteria ferment an organic material such as liquid and semi-liquid slurries, animal wastes and manures in oxygen-free conditions to produce biogas.

The main advantage of anaerobic digestion for converting waste to energy fuel is that it deals with "wet waste" which normally may be difficult to dispose off. The amount of biogas produced is limited by the size of the digester tank, so is largely used as a fuel for small-scale operations, such as farms, where enough energy can be produced to run the farm. The biogas produced can be burned in a conventional gas boiler to produce heat or as fuel in a gas engine to generate electricity or fuel some of the farm vehicles.

2.12 Gassification

2.12.1 Introduction

Gasification is used in industries for the production of electricity, ammonia and liquid fuels (oil) using Integrated Gasification Combined Cycles (IGCC), with the possibility of producing methane and hydrogen for fuel cells. Gasification is a process that converts the fossil fuel or organic waste into gases, in a gasifier. A gasifier converts solid fuel to gaseous fuel. A gasifier system includes the gasification reactor itself, along with the auxiliary equipment necessary to handle the solids, gases, and effluents going into or coming from the gasifier. The process involves oxidation and reduction of the organic matter at high temperatures. Gasification plants produce **significantly**

lower quantities of air pollutants, therefore reduce the environmental impact of waste disposal because it uses waste products as a feedstock. Gasification plants also use less water than traditional coal-based power plants.

2.12.2 Types of Gasifier

Gsifiers are classified into two categories [24]. The two categories are:

I. On the basis of the flow of oxygen in the gasifier equipment, gasifier's are classified as:

   1. Up draught gasifier
   1. Down draught gasifier
   2. Cross draught gasifier

II. On the bases of bed gasifier's are classified as:

   1. Fixed bed gasifier: Up draught gasifier, Down draught gasifier Cross draught gasifier are fixed bed gasifiers.
   2. Fluidized bed gasifier.

2.12.2.1 Up Draught Gasifier

Up draught gasifier Figure 2.18 is also known as counter current gasifier. It is the oldest and simplest type.

As seen in the diagram in updraft gasification, the fuel is supplied at the top and the air at the bottom so that fuel moves against the air flow. The gas leaves at the top. The amount of oxygen fed must be controlled as the, combustion temperature is directly proportional to it.

2.12.2.2 Down Draft Gasifier

The downdraft gasifier (Fig. 2.19) was developed to convert high volatile fuels (wood, biomass) to low tar gas and therefore has proven to be the most successful design for power generation.

- In down draft gasifier, the air/oxygen is introduced above the oxidation zone.
- The produced producer gas or synthetic gas is removed from the bottom thus, making the gas and fuel move in the same direction. Due to the fuel and gases in motion, the distilled products and acid are converted into gases like Methane, Carbon Monoxide, Hydrogen and Carbon Dioxide.

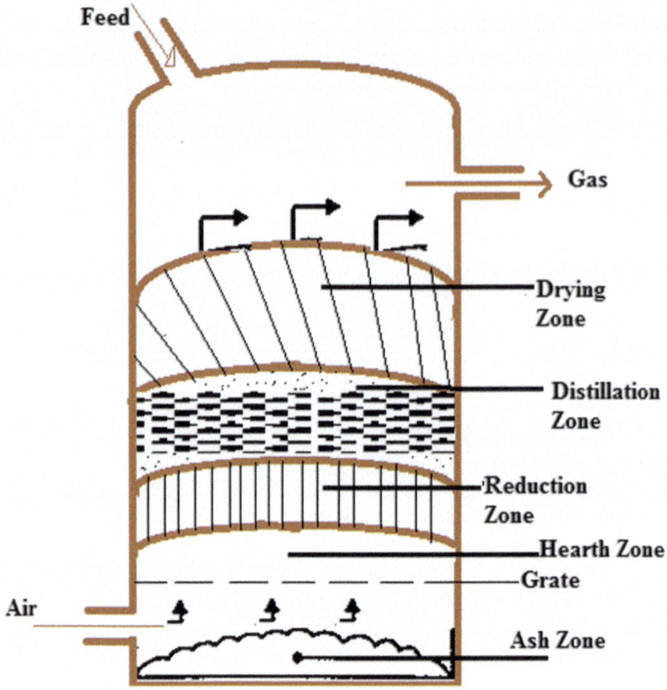

Fig. 2.18 Up draught gasifier

Advantages:

- The product obtained is the tar-free gas that can be used for engine applications.
- Due to the presence of a lower level of organic components in the condensate, these gasifiers suffer less from any environment sort of objections.

Drawbacks:

- Unable to operate on the majority of unprocessed fuels.
- They suffer from problems due to fuels with high ash content.
- There is a lack of internal heat exchange in comparison to the updraught gasifier.
- Less efficiency due to the low heating value of gases.
- The gasifier can turn inefficient under high uniform temperatures and may become impractical for power ranges above 350 kW.

2.12.2.3 Cross Draught Gasifier

Cross-draught gasifiers (Fig. 2.20) are used for the gasification of charcoal, which takes place at extremely high temperatures in the oxidation zone.

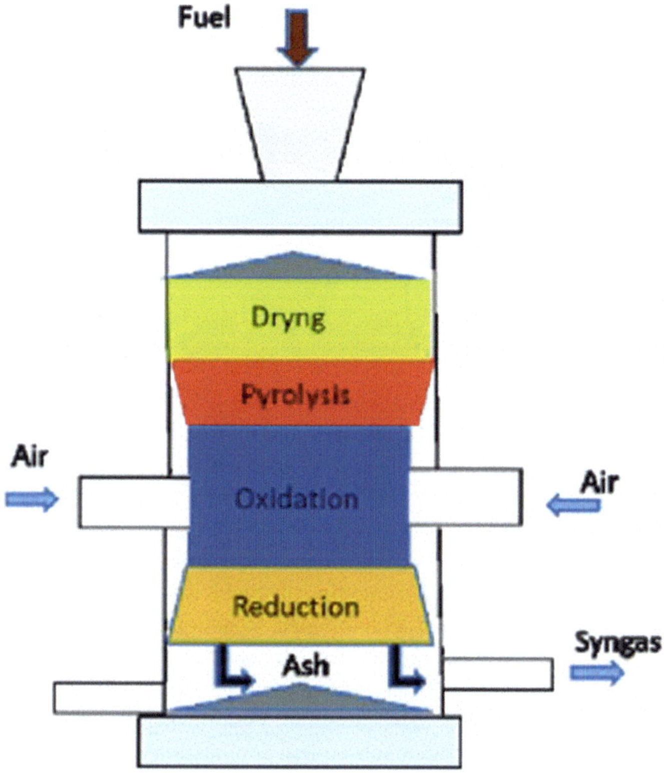

Fig. 2.19 Down draft gasifier

Fig. 2.20 Cross draft gasifier

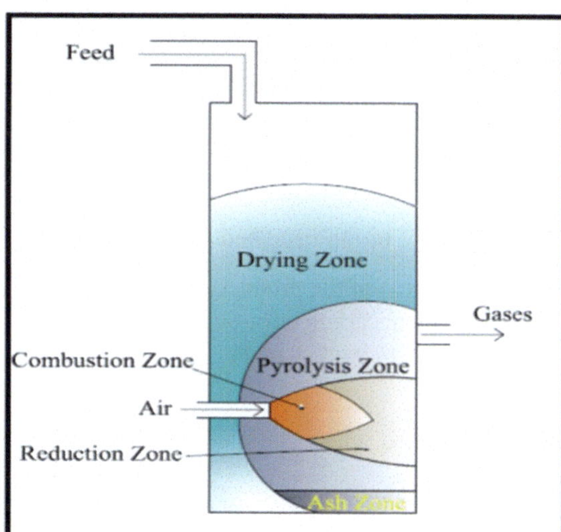

Advantages:

- The installations is economical for power range below 10 kW.
- It can be operated on a very small scale.
- These type of gasifiers can be a fit for the small engines.

Disadvantages:

- It needs a constant supply of high-quality charcoal.
- These gasifiers have minimal tar-converting capabilities.

2.12.2.4 Fluidized Bed Gasifier

Fluidized bed gasifier Fig. 2.21 has a advantage over [25] other gasifiers. The fuel is fluidized in oxygen and steam or air. Inert materials (sand or catalyst) are used as a bed material which are used for fluidization of powdery fuel and transfer of heat to it. Small particles give a better conversion because the surface area for transfer of heat is more. A fluidized bed gasifier works as a hot bed of sand particles agitated constantly by air. Air is distributed through nozzles located at the bottom of the bed. Since pre-process steps such as pelleting and briquetting the fuel is not a prerequisite, the overall operating cost is lower than other technologies. Reaction time is low compared to fixed bed gasifiers. The ash is removed dry. The temperatures are relatively low in fluidized bed gasifier, so start-up fuel requirements are lower; low-grade coals/fuels with high ash content are particularly suitable. Fuel throughput is higher than for the fixed bed gasifiers. Recycling or subsequent combustion of solids can be used to increase thermal energy recovery. Fluidized bed gasifiers are most useful for fuels that form highly corrosive ash that would damage the walls of slagging gasifiers. Biomass fuels and industrial sludge generally contain high levels of corrosive ash.

The major advantages of fluidized bed gasifiers, as reported by Van der Aarsen and others,

- Feedstock flexibility resulting from easy control of temperature,
- Ability to deal with fluffy and fine grained materials (sawdust etc.) without the need of pre-processing.

Disadvantages of fluidized bed gasifiers:

- High tar content of the product gas (up to 500 mg/m^3 gas)
- The incomplete carbon burn-out
- Poor response to load changes.

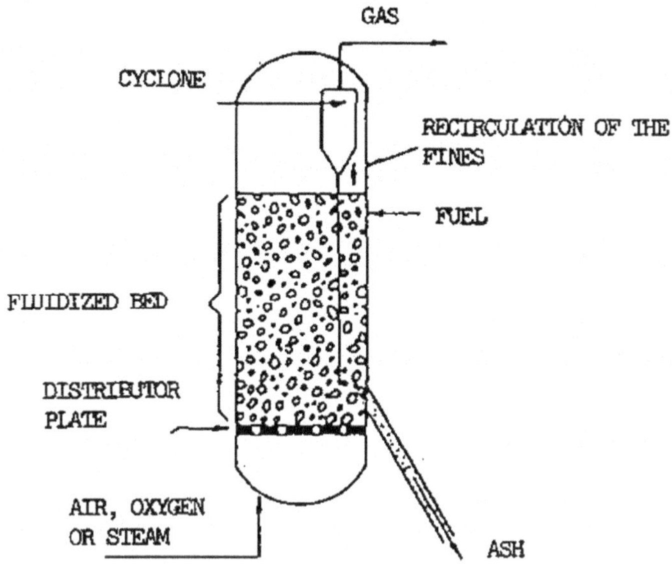

Fig. 2.21 Fluidized bed gasifier

2.13 Gasification Efficiency

The gasification efficiency determines the performance of any designed gasifier. Gassification efficiency depends on the fuel used to raise the temperature of the feed material before stimulating the production of the gas. Depending on the nature and calorific value of the feed, the gasification efficiency varies between 70 and 85%. In order to determine this parameter, major tests conducted are:

i. The determination of the elemental chemical components of the gas produced through proximate and ultimate analyses
ii. Thermal test to determine the calorific value of the produce gas using a bomb calorimeter Gasification efficiency [26] is given by the Eq. 5.2

$$\eta g = \left(\frac{(\text{Heating Value of gas} \times \text{Gas flow rate})}{(\text{Heating value of fuel} \times \text{Fuel consumption rate})} \right) \tag{2.11}$$

2.13.1 *Selection of the Right Gasifier*

Biomass fuels for gasification are many such as charcoal, wood and wood waste (branches, twigs, roots, bark, wood shavings and sawdust) and a multitude of agricultural residues (maize cobs, coconut shells, coconut husks, cereal straws, rice husks, etc.) and peat.

All the various fuels [27] differ greatly in their chemical, physical and morphological properties, therefore they require different gasification conditions resulting in the need to have different gasifier designs and thereby gasification technologies. There is no "universal" gasifier, able to handle all or most of the fuels types.

Before choosing a gasifier for any individual fuel it is important to ensure that the fuel meets the requirements of the gasifier. Practical tests are needed if the fuel has not previously been successfully gasified.

Each type of gasifier will operate satisfactorily with respect to stability, gas quality, efficiency and pressure losses only within certain ranges of the fuel properties of which the most important are:

- Energy Content
- Moisture Content
- Volatile Matter
- Ash Content
- Reactivity
- Size and Size Distribution
- Bulk Density
- Charring Properties.
- **Energy content of the fuel**

Energy content of wastes is the net calorific value of the waste. It is the heat produced by a unit quantity of waste, at constant volume and at a constant pressure of one atrnosphere.

It is assumed that all the water in the waste remains in the form of vapors. The energy content of wastes can be determined by two methods:

i. The **oxygen bomb calorimeter** is used to determine the energy content of waste under controlled conditions. The calorific value is computed from temperature observations made before and after combustion, making proper allowance for thermometer and thermo chemical corrections.

ii. **Modified Dulong Formula** [28]:

$$\text{Energy content of fuel KJ/Kg} = 337C + 1419(H_2 - 0.125O_2) + 93S + 23N \tag{2.12}$$

where:

C is the percentage of carbon,
H_2 is the percentage hydrogen,
O_2 is the percentage oxygen,
S is the percentage sulphur and
N is the percentage nitrogen.

2.13.2 *Moisture Content of the Fuel*

The moisture content of solid wastes is the weight of water in it expressed as a percentage of its wet or dry weight. Usually wet weight is commonly used.

Moisture content is defined as:

$$\text{M.Cdry} = \frac{\text{Wet Waste Weight} - \text{Dry Waste Weight}}{\text{Dry Waste Weight}} \times 100\% \tag{2.13}$$

Alternatively the moisture content on a wet basis is defined as:

$$\text{M.Cwet} = \frac{\text{Wet Waste Weight} - \text{Dry Waste Weight}}{\text{Wet Waste Weight}} \tag{2.24}$$

Conversions from one to another can be obtained by:

$$\text{M.Cdry} = [100 \times \text{M.C.wet}/100 + \text{M.C.wet}] \tag{2.15}$$

and:

$$\text{M.Cwet} = \left[100 \times \text{M.C.dry}/100 + \text{M.C.dry}\right] \tag{2.16}$$

High moisture contents reduce the thermal efficiency since heat is used to drive off the water and consequently this energy is not available for the reduction reactions and for converting thermal energy into chemical bound energy in the gas. Therefore high moisture contents result in low gas heating values.

When the gas is used for direct combustion purposes, low heating values can be tolerated and the use of feedstocks with moisture contents (dry basis) of up to 40–50% is feasible, especially when using up draught gasifiers.

In downdraught gasifiers high moisture contents give rise not only to low gas heating values, but also to low temperatures in the oxidation zone, and this can lead to insufficient tar converting capability if the gas is used for engine applications.

Both because of the gas heating value (engines need gas of at least 4200 kJ/m^3 in order to maintain a reasonable efficiency) and of the tar entrainment problem, downdraught gasifiers need reasonably dry fuels (less than 25% moisture dry basis).

2.13.3 *Volatile Matter Content of the Fuel*

The amount of volatiles in the feedstock determines the necessity of special measures (either in design of the gasifier or in the layout of the gas cleanup train) to remove tars from the product gas in engine applications.

2.13.4 Ash Content

The ash content is a measure of inorganic impurities in the fuel (typically sand, nickel, aluminium, silicon, sodium and vanadium), which can cause different kinds of problems. Ash content is determined as the weight of residue of waste after combustion in an open crucible.

2.13.5 Reactivity of the Fuel

The reactivity is an important factor determining the rate of reduction of carbon dioxide to carbon monoxide in a gasifier. Reactivity influences the reactor design insofar as it dictates the height needed in the reduction zone. Reactivity depends in the first instance on the type of fuel. For example, it has been observed that fuels such as wood, charcoal and peat are far more reactive than coal.

2.13.6 Particle Size and Size Distribution

Up and downdraught gasifiers are limited in the range of fuel size acceptable in the feed stock. Fine grained and/or fluffy feedstock may cause flow problems in the bunker section of the gasifier as well as an inadmissible pressure drop over the reduction zone and a high proportion of dust in the gas. Large pressure drops will lead to reduction of the gas load of downdraught equipment, resulting in low temperatures and tar production.

Excessively large sizes of particles or pieces give rise to reduced reactivity of the fuel, resulting in startup problems and poor gas quality, and to transport problems through the equipment. A large range in size distribution of the feedstock will generally aggravate the above phenomena.

Acceptable fuel sizes fox gasification systems depend to a certain extent on the design of the units. In general, wood gasifiers operate on wood blocks and woodchips ranging from $8 \times 4 \times 4$ cm to $1 \times 0.5 \times 0.5$ cm Charcoal gasifiers are generally fuelled by charcoal lumps ranging between $1 \times 1 \times 1$ cm and $3 \times 3 \times 3$ cm. Fluidized bed gasifiers are normally able to handle fuels with particle diameters varying between 0.1 and 20 mm.

Table 2.7 Average bulk densities

Fuel	Bulk density (kg/m^3)
Wood	300–550
Charcoal	200–300
Peat	300–400

2.13.7 Bulk Density of the Fuel

Bulk density is defined as the weight per unit volume of loosely tipped fuel. Fuels with high bulk density are advantageous because they represent a high energy-for-volume value. Consequently these fuels need less bunker space for a given refuelling time. Average bulk densities of wood, charcoal and peat are given in Table 2.7.

2.13.8 Charring Properties of the Fuel

The occurrence of physical and morphological difficulties with charcoal produced in the oxidation zone has been reported. Some feedstock's (especially softwoods) produce char that shows a tendency to disintegrate. In extreme cases this may lead to inadmissible pressure drop.

A number of tropical hardwoods (notably teak) are reported to call for long residence times in the pyrolysis zone, leading to bunker flow problems, low gas quality and tar entrainment.

2.14 Incineration

Incineration is a waste treatment process that involves the combustion of substances contained in waste materials. Industrial plants for waste incineration are commonly referred to as **waste-to-energy** facilities. Incineration of waste materials converts the waste into ash, flue gas and heat. The ash is mostly formed by the inorganic constituents of the waste and may take the form of solid lumps or particulates carried by the flue gas.

Modern incinerators reduce the volume of the original waste by 95–96%, depending upon composition and degree of recovery of materials such as metals from the ash for recycling.

Incinerators have electric efficiencies of 14–28. In order to avoid losing the rest of the energy, it can be used for e.g. district heating (cogeneration). The total efficiencies of cogeneration incinerators are typically higher than 80% (based on the lower heating value of the waste).

While incineration does not completely replace landfilling, it does significantly reduce the amount of garbage that must be disposed of. Not only that, but when utilized to treat more specialized forms of garbage, such as clinical or hazardous waste, high temperatures may remove potentially dangerous poisons and germs.

Pollutants are produced when garbage is burned. Smaller amounts of CO, NOx, HCl, HF, HBr, PCDD/F, SO_2, VOCs, PCBs, or heavy metal compounds are generated or persist after combustion, depending on the nature of the material being burnt. Most nations control these emissions and have environmental legislation in place.

The following are the various types of incinerators:

- Rotary Kiln.
- Fluidized Bed.
- Liquid Injection.
- Multiple Hearth.
- Catalytic Combustion.
- Waste-Gas Flare.
- Direct-Flame.

Rotary Kiln

Rotary kiln incineration systems are ideal for processing mixed industrial and hazardous wastes that include a combination of solid, sludge, and liquid waste streams.

Rotary kiln incinerators can be paired with custom engineered air pollution control systems for a turnkey solution to control your environment.

Rotary kiln Fig. 2.22 is a pyro-processing device used to raise materials to a high temperature (calcination) in a continuous process. Materials produced using rotary kilns include:

- Cement
- Lime
- Refractories
- Metakaolin
- Titanium dioxide
- Alumina
- Vermiculite
- Iron ore pellets.

They are also used for roasting a wide variety of sulfide ores prior to metal extraction.

Principle of Operation

The kiln is a cylindrical vessel, inclined slightly to the horizontal, which is rotated slowly about its longitudinal axis. The process feedstock is fed into the upper end of the cylinder. As the kiln rotates, material gradually moves down toward the lower end, and may undergo a certain amount of stirring and mixing. Hot gases pass along

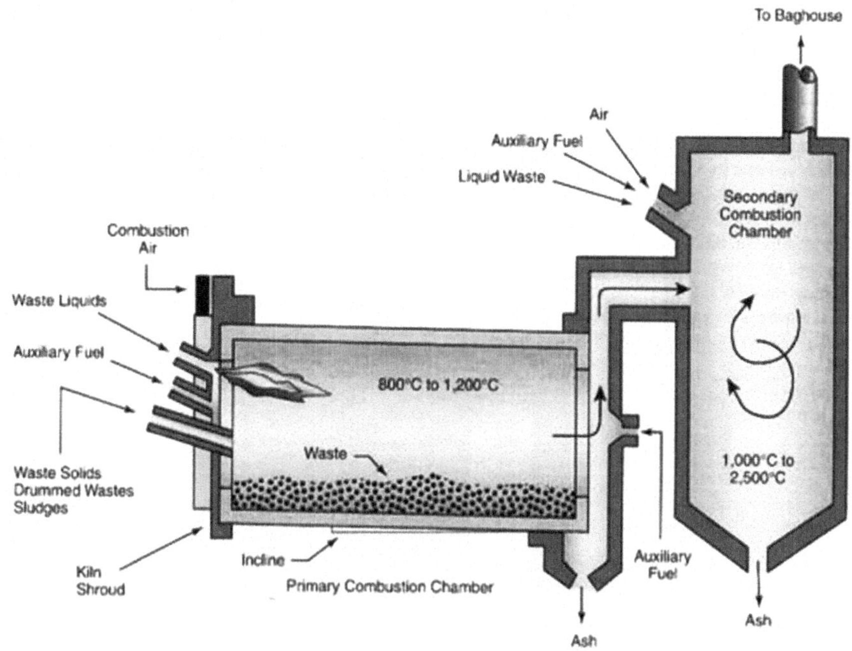

Fig. 2.22 Rotary Kiln incinerator

the kiln, sometimes in the same direction as the process material (co-current), but usually in the opposite direction (counter-current). The hot gases may be generated in an external furnace, or may be generated by a flame inside the kiln. Such a flame is projected from a burner-pipe (or "firing pipe") which acts like a large bunsen burner. The fuel for this may be gas, oil, pulverized petroleum coke or pulverized coal.

Construction

The basic components of a rotary kiln are the shell, the refractory lining, support tyres (riding rings) and rollers, drive gear and internal heat exchangers.

Advantages of Waste Incineration:

1. Reduces waste quantity
2. Reduction of pollution
3. Incinerators have filters for trapping pollutants
4. Saves on waste transportation
5. Provides better control over noise and odor in the community
6. Prevents the production of methane gas
7. Eliminates harmful germs and chemicals
8. Incineration plants can operate in any weather
9. Computerized monitoring system

10. Effective use of ash
11. Generation of heat and electrical energy.

Disadvantages of Waste Incineration:

1. Expensive
2. Pollutes the environment.

2.15 Comparison of Pyrolysis, Gasification and Incineration [29]

Parameter	Pyrolysis	Gasification	Incineration
Air supply	Total absence of oxygen	Low oxygen	Significant oxygen
Temperature range (°C)	300–800	> 750	> 850
External heat source	Required	Required	Not required
Pre-treatment of feedstock	Required (removal of glass, metals, inerts)	Required (removal of glass, metals, inerts)	Not required unless waste has organic fraction to be dried
Products	Syngas, char, tar	Syngas, char, ash	Steam, gas, ash
Scale of application	Range from small scale (30,000 tonnes/year) to large scale (500,000 tonnes/year)	Range from small scale (30,000 tonnes/year) to large scale (500,000 tonnes/year	Heat, electricity
Efficiency of energy conversion to a steam boiler	10–20%	10–20%	19–24%

References

1. https://www.prescouter.com/2017/10/waste-to-energy-technologies-available/
2. https://www.epa.gov/environmental-topics/land-waste-and-cleanup-topics
3. https://www.energy.gov/eere/renewable-energy
4. Maisaraha M et al (2018) Review on the suitability of waste for appropriate waste-to energy technology. Chem Eng Trans 63
5. Benefits of waste-to-energy solutions. https://www.plantengineering.com/articles/benefits-of-waste-to-energy-solutions/NicollePortilla December 22, 2022
6. Namratha Manohar J et al (2022) Energy content estimation from different types of wastes. In: IEEE 10th region 10 humanitarian technology conference (R10-HTC)
7. Rana R, Ganguly R, Gupta AK (2017) Evaluation of solid waste management in satellite towns of Mohali and Panchkula–India. J Solid Waste Technol Manag 43(4)

8. https://reurasia.com/advantages-of-producing-energy-from-waste
9. Maheshwari S, Deswa S (2017) Role of waste management at landfills in sustainable waste management. Int J Emerg Technol 8(1):324–328
10. https://www.ntu.edu.sg/news/detail/converting-waste-paper-into-battery-parts-for-smartphones-and-electric-vehicles
11. Advantages of the waste-to-energy system and how it helps reduce environmental impact, July 20, 2023. https://envirowaste.com.au/advantages-of-the-waste-to-energy-system-and-how-it-helps-reduce-environmental-impact/
12. https://www.giz.de/en/downloads/GIZ_WasteToEnergy_Guidelines_2017.pdf
13. Kyle Peterdy (2022) A framework to assess political, economic, social, technological, environmental, and legal factors. https://corporatefinanceinstitute.com/resources/management/pestel-analysis, November 24, 2022
14. https://www.business-to-you.com/scanning-the-environment-pestel-analysis/
15. https://web.mit.edu/urbanupgrading/urbanenvironment/sectors/solid-waste-sources.html
16. https://2ch458npc.wordpress.com/wp-content/uploads/2014/12/3-properties-of-solid-waste.pdf
17. https://www.britannica.com/technology/wastewater-treatment
18. https://zerowasteeurope.eu/2013/04/zero-waste-hierarchy/
19. https://www.researchgate.net/publication/332338946_Integration_of_Green_Energy_and_Advanced_EnergyEfficient_Technologies_for_Municipal_Wastewater_Treatment_Plants/figures?lo=1
20. https://www.fluencecorp.com/what-is-mabr/
21. https://chemicalengineeringworld.com/sedimentation-process-definition/
22. https://www.hydrotech-group.com/blog/using-roots-for-wastewater-treatment-a-root-zone-wastewater-treatment-plant-is-ideal-for-houses-production-halls-and-small-communities
23. http://www.bestbiomassmachine.com/solutions/How-to-Make-Bio-Fuel-Briquettes-with-Biomass-Briquetting-Machine.html
24. https://www.wastetoenergysystems.com/types-of-gasification-which-is-the-best-process/
25. https://www.arvindenvisol.com/fluidized-bed-gasifier/
26. Bukar AA, Oumarou MB, Tela BM, Eljummah AM (2019) Assessment of biomass gasification: a review of basic design considerations. Am J Energy Res 7(1):1–14. Available online at http://pubs.sciepub.com/ajer/7/1/1 Published by Science and Education Publishing https://doi.org/10.12691/ajer-7-1-1
27. https://www.nrel.gov/transportation/fuel-chemical-analysis-properties.html
28. Nwankwo CA etal (2016) Estimating energy content of municipal solid waste by multiple regression analysis. Int J Sci Res (IJSR) 5(6). https://www.ijsr.net/archive/v5i6/NOV164100.pdf
29. Kaza S, Bhada-Tata P (2018) Courtesy: decision maker's guides for solid waste management technologies

Chapter 3
Computation of Energy from Waste

3.1 Introduction

Any technical project can be justified only if we can quantify the input and outputs. The information is required to enable us plan the various types and quantity of input materials required, size of the plant, output energy and by products that we can get from the process, thereby know what size and type of loads we can cater to. Standards are available that provide information about the energy content of each type of waste material. Having computed the quantity of energy, we can compute the cost of the project, process, energy and thereby the financial, technical, social and environmental benefits.

3.2 Steps in Computation of Energy from Solid Waste

The output energy of the energy from waste process, commonly referred as Waste-To-Energy (WtE) can be electrical energy or biogas. The byproducts could be heat and ash. The steps for determining the electrical energy from WtE plant are:

3.1.1 Determine the quantity of each type of waste in Kilograms.

3.1.2 Determine the energy content in Kilo-Joules per Kilogram of a particular type of waste. The two methods of determined the energy are:

1. Modified Dulong Formula, 2 From Standard Tables.

3.1.3 Find the sum of the total energy in KiloJoules of all types of waste. Convert the energy in Kilojoules to KiloWattHours (KWh).

N. Manohar et al., *Machine Learning Applications in Renewable Energy*, Green Energy and Technology, https://doi.org/10.1007/978-981-97-9939-8_3

3.2.1 Determining the Quantity of Various Types of Waste Component

Determining the quantity of each type of waste component is not so easy as said. It requires high degree of planning and social commitment by all citizens of the country. The quantity of waste can be determined by two methods:

1. By segregating the waste and weighing
2. By chemical analysis of the waste samples

Determining the Quantity of waste Component by Segregation and Weighing

The primary action to determine the quantity of waste is segregation. Waste segregation is the process of sorting and separating of the various types of garbage and waste products for effective handling. Waste segregation example is the separating of various types as plastic, paper, glass, e-products for easy further management. Planning of waste segregation is the responsibility of the Municipal Administration and ensuring that each citizen follow the guidelines provided by the administration, in making waste segregation an easy task.

Waste segregation stages are:

1. Waste segregation at **Dissipater** or source end
2. Waste segregation at Landfill
3. Waste segregation at Waste-To-Energy plant.

1. **Waste Segregation at Dissipater or Source End**

Waste is any material, component or product that is of not any more use to a person, house, Institution or any organization. The person, house, Institution or any organization that does not require the material is a Waste-**Dissipater**. Thus, a waste-dissipater can be a single person or a group of persons belonging to a household, Institution or any Organization.

Most municipal administrations are providing separate bins for different types of waste. The dissipater can co-operate by following the rules and guidelines, and strictly dispose the waste in the appropriate bin. Figure 3.1 shows the various types of bins for different types of waste.

Advantages of segregation of waste at source are:

Reduces: Pollution, Litter, Toxic landfills, Costs, diseases

Fig. 3.1 Types of bins for different types of waste

Increases: Recycling, Composting, Environmental sustainability, generation of alternative energies.

2. **Waste Segregation at Landfill**

Waste segregation at landfill can effectively be done by allocation space for each type of waste. As soon as the different waste bins arrive from the dissipater, they are emptied into the allocated space. For the waste that has not been segregated, they can use tools as magnetic separators for iron metals.

Inefficiency in waste sorting results in improper disposal of waste materials for further management. Manual monitoring is inefficient, time consuming and expensive. Modern day waste segregation hardware tools have been upgraded and have become fully automatic. Some tools are smart bins. The equipment's only are not modernized but also the management strategy from waste collection to final waste usage as recycling to make a new product or combustion for generation of energy as heat and electricity is being automated using AI Techniques.

3. **Waste Segregation at Waste-To-Energy Plant**

The advancement of technology in all fields of science has made automation of waste segregation an effective and efficient process. The technology has developed several smart segregators as:

Automated Waste Sorter uses artificial intelligence for smart management f waste materials. Robots developed with artificial intelligence technique, segregate recycled waste materials from rotating conveyor belts.

RFID Tags: Waste packets are given **Radio Frequency Identification (RFID)** Tags. A pneumatically driven waste disposer, scans these tags and segregates the waste. By using ***RFID*** tags, waste can be identified and tracked through the entire process. Many waste management operations have attached RFID tags to waste containers and equipped their trucks with RFID readers thereby improving sorting, validating, verifying pickups, tracking the number of times a container is used for collection as well as tracking the weight of the container's contents at each and every pickup. A waste management system using RFID simplifies billing also. Figure 3.2 depicts some commonly used RFID tags.

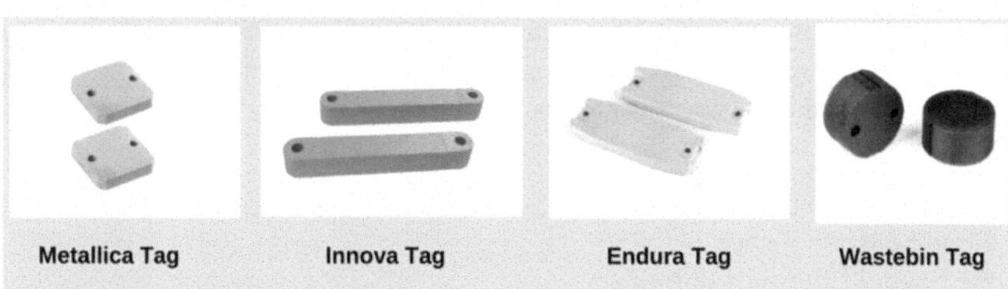

Fig. 3.2 Some commonly used RFID tags

Benefits of using RFID in Waste Management:

1. Easy to identify and sort waste.
2. Easy tracking of garbage trucks.
3. Effectual trucks movement management.
4. Meticulous disposing of hazardous and otherwise sensitive waste material.
5. Optimize collection points and containers to reduce time and cost, by adopting operation research techniques which can be easily implemented with the modern day sensors, microcontrollers and dynamic status monitoring softwares.
6. Minimum cleaning and maintenance.

IoT Sensors and Artificial Intelligence Programs: Intelligent trash can, which is built with artificial intelligence software programs and IoT detectors, is another modern development in the waste management area. The detectors on these trash cans measure and detect the garbage thrown in them and send this information to advanced servers. This system categorizes the information into the type of waste, the level, and quantity of each type of waste, and the waste disposal system, and gives a complete summary report.

After segregation of the waste the waste is weighed in appropriate weighing machine to determine he quantity in kilograms or tonnes.

Determining the quantity of waste by chemical analysis of the waste samples.

In cases where all types of waste have been mixed, a sample of the waste is collected, and by conducting chemical analysis in the laboratory the various components or types of waste are determined.

3.2.2 Determine the Energy Content in Kilo-Joules Per Kilogram of a Particular Type of Waste

Energy Content: Energy content, of solid wastes is the net calorific value of waste material component. It is the heat produced by a unit quantity of waste, at constant volume and at a constant pressure of one atmosphere. The assumption is that all the water in the waste is in the form of vapour. The energy content of municipal solid wastes can be determined by means of an oxygen bomb calorimeter under controlled conditions. The calorific value is computed from temperature observations made before and after combustion, making proper allowance for measuring instruments as thermometer and thermo chemical corrections.

Three methods of determined the energy content are:

1. Modified Dulong Formula
2. From Standard Tables
3. R1 Energy Efficiency Formula

1. **Modified Dulong Formula**

$$\text{Energy content of fuel KJ/Kg} = 337\text{C} + 1419(\text{H}_2 - 0.125\text{O}_2) + 93\text{S} + 23\text{N}$$

where:

C percentage carbon,
H_2 percentage hydrogen,
O_2 percentage oxygen,
S percentage sulphur and.
N percentage nitrogen.

2. **From Standard Tables**

The standard tables provide the energy content in KiloJoules per Kilogram of each type of waste.

Energy content of a particular type of waste (KiloJoules) = [Total weight of the fuel in Kg × Energy content in KiloJoules/Kg]

3. **R1 Energy Efficiency Formula**

The **"R1- Energy Efficiency formula"** is a performance indicator for the level of recovery of energy from waste in a plant dedicated to the incineration of municipal solid waste (MSWI).

WFD (Waste Framework Directive 2008/98/EC) has scheduled the formula used to calculate the value of energy efficiency. The "R1 Energy Efficiency Formula" is:

$$\text{Energy Efficiency} = \frac{(Ep - (Ef - Ei)}{(0.97X(Ew + Ef))}$$

where:

E*p* Annual energy produced as heat or electricity. = (Energy in the form of electricity × 2.6) (GJ/year)
E*f* Annual energy input by the fuel resulting in production of steam(GJ/year)
E*i* Annual energy imported excluding E*w* and E*f*(GJ/year)
E*w* Annual energy contained in the treated waste. It is computed using the net calorific value of the waste (GJ/year)
0.97 The factor for energy losses due to bottom ash and radiation.

R1 formula:

i. Computation results in the efficiency at which the produced energy is utilized.
ii. It can be applied to incineration operation process that meet the following criteria:

- Energy by combustion of waste must be more than energy consumed by the process itself.
- Large quantity of the waste must be consumed during the operation.
- Efficiency of the process should be maximum, to obtain optimum output in the form of heat or electricity.

3.2.3 Converting the Energy in Kilojoules to Kilowatt-Hours (KWh)

i. Energy contained in Total waste (KiloJoules) = $\sum$(Energy contained in all types of waste in KiloJoules).
ii. Convert the total KiloJoules obtained in above step to KiloWatt-hours,using the conversion factor:

$$1\ \text{KiloJoule} = 0.0027778\ \text{KWh}$$

3.3 Solved Problems

3.3.1 Solid Waste Management

1. **Determine the area of landfill for waste required for a population of 75,000, given the following data?**

   (i) Solid waste generation per day: 1.55 kg/person
   (ii) Compacted density of solid waste in landfill: 510 kg/m^3
   (iii) Average compacted depth of solid waste: 3.5 M

   Solution:
   Figure 3.3 depicts a landfill

   Total population = 75,000
   Solid waste generation: 1.55 kg/person/day
   Compacted density of solid waste in landfill: 510 kg/m^3
   Average compacted depth of solid waste: 3.5 M
   Total Solid Waste generated (Mass) = Total population × solid waste generated/person/day

$$75{,}000 \times 1.55 = 1,16,250\ \text{Kg}$$

$$\text{Volume of Total Solid Waste} = \text{Mass/Density} = 116250/510 = 228\text{m}^3$$

$$\text{Landfill Area required} = \text{Volume/Depth} = 228/3.5 = 65.13\text{m}^2$$

   a. **Determine the landfill area required for municipality with a population of 100,000, given that solid waste generation is = 950 gm/person/day, compacted density of landfill = 505 kg/m^3.Average depth of compacted solid waste = 4 m.**

2. **A colony having a population of 65,000 generates solid wastes at the rate of 1 kg/person/day. The compacted specific weight of solid wastes in landfill**

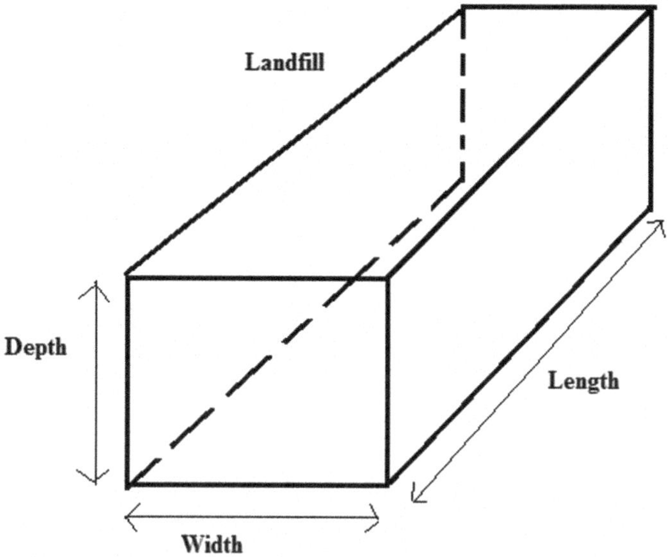

Fig. 3.3 Landfill

is 650 kg/m³ and the average depth of compacted solid wastes in landfill is 5 M. Determine the required landfill?

Solutions:

Total population = 65,000 people
Solid waste generated per person per day = 1 kg
Density of solid waste = 650 kg/**m³**
Depth of landfill = 5M
Total solid wastes generated from the colony, per day = 65,000 × 1 = 65,000 kg/day
Volume of solid wastes per day = Total mass / density = 65,000/650 = 100m²/day
Area required = Volume / depth = 100/5 = 20m²/day
Area required annually = 20 × 365 = 7300 m²/year

3. **Estimate the unit solid waste generation rate for a residential area having 950 dwellings. The observation made at local transfer station for a week time revealed that the waste was carried in self—compacting trucks and flatbed trucks, who's volume is 18 m³ and 2 m³and the density of waste is 310 kg/m³ and 120kg/m³respectively. Assume 15 self- compacting truck and 16 flatbed truck loads per week.**

Solution:

Total No. of dwellings = 950

Self—compacting trucks: No. = 15
Volume = $18 m^3$
Density of waste = 310 kg/m^3
Flatbed trucks: No. = 16
Volume = $2 m^3$
Density of waste = 120 kg/m^3

$$\begin{aligned}\text{Total weight of solid waste generated} &= \text{No of trucks} \times \text{Volume} \times \text{density} \\ &= [15 \times 18 \times 310] + [16 \times 2 \times 120] \\ &= 83700 + 3840 \\ &= 87,540\,\text{Kg}\end{aligned}$$

Unit solid waste generation rate per week per dwelling = Total solid waste generated/No. of dwellings

$$87540/950 = 92.15\,\text{Kg}$$

Unit solid waste generation rate per day per dwelling =
Unit solid waste generation rate per week per dwelling/7

$$= 92.15/7 = 13.16\,\text{Kg}$$

4. **The total number of students in a high school is 900. The school has 31 classrooms. Assuming a 6-day school week with solid waste being picked up on Wednesday and Saturday before school starts in the morning; determine the size of storage containers required. Assume waste generation rate = 0.12 kg/capita/day and 4.2 kg/room, density of waste = 125 kg/m^3, standard containers are 1.5, 2.5, 3.0, and 4.5 m^3**

Solution:

Population of a high school = 900
No. of Class-Rooms = 31
Waste Generation Rate = 0.12 kg/capita/day and 4.2 kg/room
Density of waste = 125 kg/m^3
Total waste production = [Total No. of students × Waste generation rate] + [No. of Class rooms × waste generated per room]

$$= \left[900 \times 0.12\right] + [31 \times 4.2] = 108 + 130.2 = 238.2\,\text{kg/day}$$

Total volume of waste = Mass/Density = 238.2/125 = 1.906 m^3/day
Since the maximum storage period is 3 days (Wednesday, Thursday and Friday), we have
Volume of waste generated in 3 days = 3 days * 1.906 m^3/d = 5.717 m^3

The best solution to this would be: 1 container of 1.5 m^3 and 1 container of 4.5 m^3.

5. Estimate the unit solid waste generation rate of a residential area with 1500 houses and 6 people in each house. A week-long observation at local single-use stations revealed:

Vehicle	Number of loads	Average volume (m^3)	Density kg/m^3
Truck	14	10	350
Tractor	7	2	150
Private vehicle	20	1	100

Solution:

Vehicle (1)	Number of loads/ week (2)	Average Volume (m^3) (3)	Density kg/m^3 (4)	Mass of solid waste generated Kg/week (5 = 2 * 3 * 4)
Truck	14	10	350	49,000
Tractor	7	2	150	2100
Private vehicle	20	1	100	2000
			Total solid waste	= 53100 kg

Total Solid Waste/Day = 53,100 /7 = 7586 kg
Unit solid waste generation per person = Total solid per day/(No. of houses × No. of persons per house) = 7586/(1500 × 6) = 0.843 kg

6. In a residential area with 2000 households, there are an average of 5 residents per household. In order to estimate the amount of solid waste generated, the following observations were made at the waste disposal site over a period of one week:

Type of vehicle	No. of trips	Volume (m^3)	Specific weight (kg/m^3)
I	10	15	300
II	15	2	150
III	30	1.0	100

Determine the unit rate of solid wastes generation.

Solution:

Type of vehicle (1)	No. of trips (2)	Volume (m^3) (3)	Specific Weight (kg/m^3) (4)	Total weight (Kg) Col: 2 * 3 * 4
I	10	15	300	45,000
II	15	2	150	4500
III	30	1.0	100	3000
Total Kg/week				52,500

Unit Generation Rate per person per day = (Total Waste Generated)/(No.of Houses * Residents per house * days)

$$52500/(2000 * 5 * 7) = 0.75\,\text{kg/capita/day}$$

7. **Estimate the density of solid waste sample on as discarded basis. Take 1000 Kg sample.**

Component	% by mass	Density (Kg/m^3)
Food waste	20	300
Paper	40	100
Plastics	5	90
Garden trimmings	15	150
Wood	5	250
Tin cans	5	100
Cardboard	10	80

Solution:

Component (1)	% by mass (2)	Density (Kg/m^3) (3)	Mass in Kg (4) 1000 × (Col.2) /100	Volume m^3 Density/Mass (Col.3/Col.4)
Food waste	20	300	200	1.50
Paper	40	100	400	0.25
Plastics	5	90	50	1.80
Garden trimmings	15	150	150	1.00
Wood	5	250	50	5.00
Tin cans	5	100	50	2.00
Cardboard	10	80	100	0.80
Total mass Kg			1000	12.35

Density of Total waste = Total Mass/Total Volume = 1000/12.35 = 80.97 kg/m^3

8. Estimate the moisture content of MSW sample with following Composition

Component	% by mass	Moisture content %
Food waste	25	70
Paper	40	6
Cardboard	10	5
Plastic	10	2
Garden trimings	10	60
Tin cans	5	3

Solution:
Considering Total MSW of 1000 kg:

Component (1)	% by mass (2)	Moisture content % (3)	Weight of component in Kg (4 = Col.2/100 × 1000)	Weight of dry mass in Kg (5 = [(100 – Col.3) /100] × Col.4
Food waste	25	70	250	75
Paper	40	6	400	376
Cardboard	10	5	100	95
Plastic	10	2	100	98
Garden trimings	10	60	100	40
Tin cans	5	3	50	48.5
		Total	1000 kg	732.5

$$\begin{aligned}\text{Moisture Content} &= (\text{Total Weight} - \text{Weight of dry mass})/\text{Total Weight}\\ &= (1000 - 732.5)\\ &= 267.5\ \text{Kg}\end{aligned}$$

$$\%\ \text{Moisture Content} = (267.5 * 100)/1000 = 26.75\%$$

9. **Calculate the energy content of a solid waste sample with the following composition?**

Component	% by mass	Energy KJ/Kg
Paper	40	17,770
Cardboard	12	15,800

(continued)

(continued)

Component	% by mass	Energy KJ/Kg
Plastics	12	16,400
Food wastes	16	5000
Garden wastes	8	7000
Wood	6	19,000
Tin cans	6	600

Solution:
Assume 100 kg of sample.

Component	% by mass	Energy KJ/kg	Total Energy (KJ)
(1)	(2)	(3)	(4) (2 * 3)
Paper	40	17,770	710,800
Cardboard	12	15,800	189,600
Plastics	12	16,400	196,800
Good wastes	16	5000	80,000
Garden wastes	8	7000	56,000
Wood	6	19,000	114,000
Tin cans	6	600	3600
–	100%	–	1,350,800 kJ

Unit Energy Content = 1,350,800/100 = 13508 kJ/kg.

10. **Calculate the energy content of the solid waste sample having following composition using modified Dulong equation.**

Component	% by mass
Carbon	35
Hydrogen	10
Oxygen	40
Nitrogen	8
Sulphur	3
Ash	4

Dulong equation

$$\begin{aligned} \text{Energy Content}(\text{KJ}/\text{Kg}) &= 337.2\text{C} + 1418(\text{H} - \text{O}/8) + 93\text{S} + 23\text{N} \\ &= 337 \times 35 + 1418 \times (10 - 40)/8 + 93 \times 3 + 23 \times 8 \\ &= 11795 - 5317.5 + 279 + 184 \\ &= 6940.50\ \text{KJ}/\text{Kg} \end{aligned}$$

11. **Estimate the i. moisture content, ii. density and iii. energy content of a solid waste sample that has following components. Use suitable data**

Component	% by mass
Food waste	20
Paper	35
Cardboard	10
Plastics	15
Yard wastes	10
Wood	5
Tin cans	5

Solution:

i. Moisture Content:

Considering 100 Kg of Waste
Considering Moisture Content% from Annexure-I

Component (1)	% by mass (2)	Mass in Kg (3) = (Col.2 × 100/100)	% Moisture from standard Tables (4)	% Dry waste (5 = 100 – Col.4)	Dry mass(Kg) (5 = Col.5/100 × Col.3)
Food waste	20	20	70	30	6.00
Paper	35	35	6	94	32.90
Cardboard	10	10	5	95	9.50
Plastics	15	15	2	98	14.70
Yard wastes	10	10	60	40	4.00
Wood	5	5	20	80	4.00
Tin cans	5	5	3	97	4.85
				Total	75.95

Moisture Content in Total Waste = (Total Weight – Dry Weight)/Total Weight
= [(100 – 75.95)/100] × 100 = 24.05%

ii. Density:

Considering Density from Annexure-II, the volume corresponding to 1000 kg of sample is computed as follows:

Component (1)	% by mass (2)	Density in Kg/m^3 from standard tables (3)	Mass in 1000 kg (4 = 1000 × Col.2/100)	Volume(m^3) (5 = Col.4/Col.3)
Food waste	20	290	200	0.69
Paper	35	85	350	4.12

(continued)

(continued)

Component (1)	% by mass (2)	Density in Kg/m^3 from standard tables (3)	Mass in 1000 kg (4 = 1000 × Col.2/100)	Volume(m^3) (5 = Col.4/Col.3)
Cardboard	10	50	100	2.00
Plastics	15	65	150	2.31
Yard wastes	10	105	100	0.95
Wood	5	240	50	0.21
Tin cans	5	90	50	0.56
			Total	10.84

Density = 1000/10.84 = 92.25 Kg/m^3

iii. Energy Content:

Considering Energy Content from Annexure-III, the total energy corresponding to 100 kg of sample is computed as follows:

Component (1)	% by mass (2)	Energy content KJ/ Kg from standard tables (3)	Total energy (KJ) (4 = 100 × Col.2/100 × Col.3)
Food waste	20	4650	93,000
Paper	35	16,750	586,250
Cardboard	10	16,300	163,000
Plastics	15	32,600	489,000
Yard wastes	10	6500	65,000
Wood	5	18,600	93,000
Tin cans	5	700	3500
			Total = 1,492,750

Source https://www.chegg.com/homework-help/questions-and-answers/problem-1-energy-content-dry-basis-ash-free-basis-solid-waste-sample-composition-given--as-q55549283#question-transcript

Total Energy = 1,492,750KJ
Since, 1 kJ = 0.000278 kWh
Total Electrical Energy = 414.652778 kWh
Unit Energy Content = 1492750/100 = 14927.50 KJ/Kg (as discarded)

12. **Estimate the energy content of a solid waste sample with the following composition:**

Assume moisture content = 21%
Ash Content = 5%
What is the energy content on dry basis and on ash free dry basis? Use suitable data

Component	% by mass
Food waste	15
Paper	45
Cardboard	10
Plastics	10
Yard wastes	10
Wood	5
Tin cans	5

Solution:

Considering Energy Content from Annexure-III, the total energy corresponding to 100 kg of sample is computed as follows:

Component (1)	% by mass(2)	Energy content KJ/ Kg from standard tables (3)	Total energy (KJ) (4 = 100 × Col.2/100 × Col.3)
Food waste	15	4,650	69,750
Paper	45	16,750	753,750
Cardboard	10	16,300	163,000
Plastics	10	32,600	326,000
Yard wastes	10	6500	65,000
Wood	5	18,600	93,000
Tin cans	5	700	3500
Total energy content (KJ)			1,474,000

$$\text{Energy content} = 1,474,000/100 = 14,740\,\text{KJ/Kg}$$

$$\text{EC on dry basis} := \textit{Total Weight} \times \left[\frac{\textit{Total Weight}}{(\textit{Total Weight} - \%\,\textit{Moisture Content})}\right]$$
$$= 14740 \times [100/(100 - 21)] = 18,658\,\text{KJ/Kg}$$

EC on ash free basis

$$= \textit{Total Weight} \times \left[\frac{\textit{Total Weight}}{(\textit{Total Weight} - \%\textit{Moisture Content} - \%\textit{Ash Content}}\right]$$
$$= 14,740 \times [100/(100 - 5 - 21)] = 19919\,\text{KJ/Kg}$$

13. **A comingled MSW having particles of average size 300 mm, is to be reduced to final size of 50 mm in a plant having a capacity of 100Ton/hr. Determine the energy required by the plant if a specific energy of 20hp-h/ton is required to reduce MSW of size 15–50 mm.**

Solution:
The Energy consumption can be calculated using Kick's Law,

$$\textbf{Kick's Law} = E = Cln\left(\frac{l1}{l2}\right)$$

The Calibration constant C can be calculated using 20hp-h/ton is required to reduce MSW of size 15–50 mm.

So E = 20, l_1 = 150 mm, l_2= 50 mm

$$20 = Cln\left(\frac{150}{50}\right)$$

C = 18.2hp-h/ton

Energy Consumption to reduce from 300mm to 50mm is

$$E = 18.2ln(300/50) = 32.618\,\textbf{hp/ton}$$

The energy required by the plant = 100 x 32.618 = 3261.8hp = 0.746 x 3261.8 = 2433.3KW.

That is about 2.4 MW.

3.4 Tutorial Problems

(1) A community of 1000 homes stores their waste containers along alleys located in back of all of their homes. A satellite vehicle with a 2-yd3 capacity picks up the waste once per week and delivers it to a mechanically loaded compactor truck that moves along a main artery as the satellite vehicle collects wastes. Calculate the number of trips per week the satellite vehicle makes to the compactor truck and the time per week required to collect all of the waste. How many trips does the compactor make to the disposal site?

(2) Solid waste is collected from a locality using a stationary container system, provided with two containers each of size 4 cubic meters and utilization factor of 0.75, at each location. The truck takes 20 min to reach to first container from garage. Six minutes are spent for unloading each container into the truck and average time taken by the truck to cover the distance between consecutive container locations is also 6 min, the round trip haul distance is 60 km and the truck takes 15 min to reach the garage from last container location. Estimate the required capacity of the truck if the number of trips to disposal site per day is 2 (8 h workday). Speed limit of truck is 40 kmph and collection vehicle compaction ratio is 2.5.

(3) Calculate the energy values of MSW collected from different countries by using the modified Dulong formula?

Component	% by mass
Carbon	65
Hydrogen	8
Oxygen	20
Nitrogen	2
Sulphur	1
Ash	4

(4) Calculate the energy content of the solid waste having following composition using modified Dulong equation.

Component	% by mass
Carbon	36.3
Hydrogen	7.3
Oxygen	51.1
Nitrogen	0.5
Sulphur	0.1
Ash	4.7

3.5 Biogas Plant Energy

(1) A cow produces an average of 12 kg of manure per day. Suppose a family has 5 cows. Calculate the capacity of the biogas plant to be built?

Solution:

Cow Dung/day = 5 cow $\times$ 12 kg/cow/day = 60 kg of cow dung/day
Since 1 kg of cow dung yields 0.04 m^3 or 40 litres of biogas
60 kg will produce 60 $\times$ 40 = 2400 litres or 2.4 m^3 of bio-gas in a day.
Therefore, the capacity of the plant to be constructed should be 2.5 m^3.

(2) If 0.65 cubic metres of biogas produces 1 kW of biogas per hour, how much biogas is required per month for a 5 kW generator running five hours per day?

Solution:

$$\text{Biogas required per month} = (0.65\,\text{m}^3/\text{kW/h}) \times 5\,\text{hours} \times 5\,\text{kW}$$
$$= 16.25\,\text{m}^3$$

(3) **What should the daily input of slurry be for a 5 m³ (175 cu ft) biogas plant using cattle dung?**

Solution:
For 5 m³ of daily gas production cattle dung required = 5 x 32 = 160 kg = 160 litres (1 kg dung equals approximately 1 litre volume)
Considering slurry ratio of 1:1 =
Daily slurry input = 160 × 2 = 320 litres

(4) He has to cook for five members of his family and he expects that number to increase to seven in the near future. He wants 3 lamps, one for 3 h a day, one for 4 h a day, and one for 2 h a day. Estimate the size of the biogas plant he requires to install?

Solution:
Total Production:

Animal (1)	No. of animals (2)	Excreta (Kg)/ animal (3)	Energy litres of gas/kg (4)	Total excreta (Kg) (5 = 2 × 3)	Total energy m^3 (6 = 4 × 5)
Pig	26	3	50	78	3.900
Cow	4	10	37	40	1.5
				Total energy = 5.4 m^3	

Total Energy Required:

Activity	No.	Standard energy required	Total
Cooking	7 People	0.3 m^3/person/day	$7 \times 0.3 = 2.1 m^3$
Lighting	3 lights, one for 3 h per day, one for 4 h per day and one for 2 h per day	0.150 m^3/hour	$(3 + 4 + 2) \times 150 = 1.35 m^3$
		Total energy = 3.45 m^3	

Standard sizes of plant available are 3 m^3 or 5 m^3/day.

It would be better for him to install 5 m^3 Biogas Plant, as it would support future expansion also.

(5) Given the following data compute the total biogas produced in a month:

Types of dung	Gas production M^3 per Kg dung (mJ)	Dung available/day
Cattle (cows and buffaloes)	0.035	50
Pig	0.055	30
Poultry (chickens)	0.110	10

Compute the total KWh given 1 M^3 of biogas = 2KWh.

Types of dung	Gas production per Kg Dung (mJ)	Dung available/day	Total gas/day
Cattle (cows and buffaloes)	0.035	50	1.75
Pig	0.055	30	1.65
Poultry (Chickens)	0.110	10	1.10
		Total gas/day	4.50 M^3

Total Biogas produced in a month = Biogas/day × 30 = 4.5 × 30 = 135.0 M^3
Total KWh = 135 × 2 = 270 KWh.
Total Electrical Energy /Month = 270 KWh.

(6) There are 350 cows and 150 calves in a dairy farm. Each dairy cow produces an average of 12 kg of cow dung per day. Compute i. the total biogas produced in a month ii. Electrical Energy generated in a month and iii. Volume of the biogas plant

Given: Average biogas/kg of mixed gobar per day = 0.04 m^3
Average bulk density of gobar = 920 kg/m^3
1 m^3 of biogas = 2KWh

Solution:

Number of cows = 350
Number of calves = 150.So, 75 adult cows are assumed)
Therefore total number of cows = 350 + 75 = 425
Average gobar produced by 425 cows = 425 × 12 = 5100 kg/day
Average biogas available per day = 5100 × 0.04 = 204 m^3

Total biogas produced in a month = 204 × 30 = 6120 m^3

(ii) Electrical energy per month = 6120 × 2 = 12240 KWh
(iii) Volume of the digester = Mass per day/Density = 5100 / 920 = 5.54 m^3

Therefore, a digester of 6 m^3 would be optimum.

(7) **If daily slurry supply is 80 litres and Retention time is 90 days then determine the digester volume?**

Solution:

Daily slurry supply = 80 L
Retention time = 90 days
Digester volume (V_d) = Daily slurry supply * Retention time
= 80 l/day × 90 days = 7200 1 (7.2 m^3)

(8) **If Digester volume (V_d) is 4500 l and Daily supply (S_d) is 50 l/day, what is the retention time? Digester volume (Vd) = 4500l**

Solution:

Daily slurry supply = 50 l
Retention time = Digester volume (Vd) / Daily slurry supply
= 4500/50 = 90 days

(9) **If Digester volume (V_d) is 5000 l and Retention time is 60 days, then determine the Daily slurry supply required?**

Solution:

Digester volume (V_d) = 5000 l
Retention time = 60 days
Daily slurry supply = Digester volume (V_d)/Retention time
= 5000 × 60 = 300,000 l = 300Kl

(10) **Calculate the following parameters of a biogas system.**

(i) **Volume/Capacity of Biogas Digester.**
(ii) **Power generated from biogas Digester**

Given:
Calorific value of methane: 28 mJ/m^3
Burner Efficiency: 70%
No. of cows: 8
Retention period: 20 days
Temperature of fermentation: 300 °C
Cow dung collected per cow per day: 2 kg
Density of slurry in digester = 50 kg/m^3
Biogas yield=0.2 m^3/kg of cow dung
Methane proportion in Biogas: 0.7
Solution:
Given:
No. of cows = 8; Burner Efficiency = 0.7
Calorific value of methane, H_m: 28 mJ/m^3
Retention Period = 20 Days.
Dry matter (cow dung) Collected /cow /day = 2 kg
Density 50 kg/m^3
Biogas Yield = 0.2m^3/kg of dry input
Methane proportion in biogas = 0.7

(i) **Volume of biogas digester:**

Mass of dry input/day M = dry matter/cow/day × no. of cows
= 2 × 8 = 16 kg /day
Volume of fluid in digester V_f = (Mass in kg/day)/(Density in Kg/m^3)
= 16/50 = 0.32m^3/day

Digester volume $V_d = V_f \times$ no. of days of retention period
$= 0.32 \times 20 = 6.4\ m^3$
Volume of biogas $V_b =$ Biogas Yield Per Kg of dry Input $\times$ Mass of dry Input
$= 0.2 \times 16$
$= 3.2 m^3/day$

(ii) **Power Available from digester E:**

E = Efficiency of burner $\times$ C.V. of methane $\times$ Volume of biogas (V_b) $\times$ factor of methane in Gas
$= 0.7 \times (28 \times 10^3\ KJ/m^3) \times 3.2 \times 0.7$
$= 43{,}904$ KJ/day $= 43{,}904 \div (24 \times 3600)$
E = 0.508 KW

(11) **How much gas will be required per day for an incubator which measures 50 cm x 50 cm x 30 cm inside?**

Solution:
Volume of incubator $= (50 \times 50 \times 30)/1{,}000{,}000 = 0.075\ m^3$
Since $0.7\ m^3$ gas per hour per m^3 incubator capacity is needed, for the present incubator
Gas required/day $= 0.075 \times 0.7 \times 24 = 1.26\ m^3$

(12) **How much biogas will be required per day to operate a small refrigerator of 125 litres ($0.125\ m^3$) capacity?**

Solution:
Gas required per day $= 0.125 \times 1.2 \times 24 = 3.6\ m^3$
(Since $1.2\ m^3$ per hour per m^3 refrigerator capacity is needed).

(13) **Compute the amount of biogas required to run a 5kW generator for three hours each day?**

Solution:
Average BioGas Required per KW/hr $= 0.65\ m^3$
Therefore gas required for 5 KW generator to operate 3 h /day.
$= 5 \times 3 \times 0.65 = 9.75\ m^3 \approx = 10\ m^3$

(14) Consider the following data of a domestic biogas plant:

Number of cows = 10
Retention Time = 20days
Temperature = 30 °C
Dry matter/cow = 2 Kg/day
Biogas yield $= 0.25\ m^3/kg/cow$
Burner Efficiency = 60%
Methane proportion = 0.8
Heat of combustion of methane $= 28\ MJ/m^3$
Density of dry material $= 50\ Kg/m^3$
Determine the power available from the digester.

Total mass of dry matter = Number of cows × Dry matter/cow
= 10 × 2 = 20 kg/day
Volume of biogas = (Biogas yield × Total mass of dry matter consumed)
= 0.25 × 20 = 5.0 m^3/day
Power available from the digester = (Efficiency of burner × Heat of combustion of methane × Methane proportion × Volume of biogas).
= 0.6 × 28 × 0.8 × 5 = 40.32 MJ/day.

(15) **Calculate the following parameters of a biogas system. (a) Volume of Biogas Digester. (b) Power available from biogas Digester**

Given:
Calorific value of methane: 28mJ/m^3
Burner Efficiency: 70%
No. of cows: 8
Retention period: 20 days
Temperature of fermentation: 300 °C
Dry matter (cow dung) collected per cow per day: 2 kg
Density of matter in fluid (slurry) in digester = 50 kg/m^3
Biogas yield = 0.2 m^3/kg of dry input
Methane proportion in Biogas: 0.7
Solution:

(1) **Volume of biogas digesters:**

Mass of dry input/day M = dry matter/cow/day × no. of cows
= 2 × 8 = 16 kg /day
Volume of fluid in digester, Vf = m(kg/day)/(P(Kg/m^3)
= 16/50 = 0.32 m^3/day
Digester volume Vd = Vf × no. of days of retention period
= 0.32 × 20 = 6.4 m^3
Volume of biogas Vb = Biogas Yield Per Kg of dry Input × Mass of dry Input
= 0.2 × 16
= 3.2m^3/day

(2) Power Available from Digester E:

E = Efficiency of burner × C.V. of methane × Volume of biogas (Vb) × factor of methane in Gas
= 0.7 × (28 × 10^3 kJ/m^3) × 3.2 × 0.7
= 43,904 kJ/day = 43,904 ÷ (24 × 3600)
E = 0.508 Kw

Formulae

- Density = Mass/volume
- Volume of landfill = length × breadth × height
- Total Solid Waste generated = Total population × solid waste generated/person/day

- Landfill Area required = Volume/Depth
- **Dulong equation**: Energy Content (KJ/Kg) = 337.2C + 1418 (H–O/8) + 93S + 23N
- **Kick's Law:**

The amount of energy required to crush a given quantity of material to a specified fraction of its original size is the same no matter what the original size.

$$E = Cln\left(\frac{l1}{l2}\right)$$

where:

E Energy Consumption rate, hp-h/ton
C Calibration constant, hp-h/ton
l_1 Initial size
l_2 Final size

Formula to calculate the volume of a digester and the power available from the digester:

i. Mass of dry input/day M or m_o = dry matter/cow/day × no. of cows
ii. Volume of fluid in digester, Vf = mass / density = m_o (kg/day)/(ρ(Kg/m^3)) [ρ = 50 kg/m^3]
iii. Digester volume Vd = Vf × no. of days of retention period
iv. **Power Available from digester E** = Efficiency of burner × Heat of combustion of methane × Volume of biogas (Vb) × factor of methane in Gas

$$\mathbf{E} = n * \mathbf{H}_m * Vb * F_m$$

- For any one feedstock, daily biogas production can be estimated using the following equation (Fulford 2015):

$$G = C \times Vd \times S \times \left(\frac{k}{1 + kR}\right)$$

where: —G is the biogas production (in m^3/day)

- C is the biogas potential, which is the maximum amount of gas that can be produced from 1 kg of volatile solids in a feedstock (in m^3/kg).
- Vd is the digester volume.
- S is the initial concentration of volatile solids in the slurry (in kg/m^3).
- R is the feedstock retention time (in days).
- k is a constant indicating the rate of gas production at a given temperature.

IRENA has simplified it as

$$G = \left(Y \times Vd \times \frac{S}{1000} \right)$$

where Y is a yield factor based on temperature and the feedstock retention time

- When wastes are compressed, their volume is reduced, which is normally expressed in percentage, given below:
- Volume Reduction (%) = [Vi − Vf/Vi] × 100
- The compaction ratio of the waste is given in equation
- Compaction ratio = Vi/V f
- where Vi = volume of waste before compaction, m^3 and
- V_f = volume of waste after compaction, m^3

3.5.1 Conversions

1 MJ = 0.278 KWh
$1M^3$ = 2 KWh
1KWh = 860Kilo Calories of Thermal Energy = 3600 kilo Joules of thermal energy

Annexure-I

Typical Data of Moisture Content and Ash Content of Municipal Solid Waste (MSW) Components

S. No.	Component	Moisture %	Ash%
1	Food waste	70	5
2	Paper	6	6
3	Cardboard	5	6
4	Plastic	2	10
5	Textile	10	2.5
6	Rubber	2	10
7	Leather	10	10
8	Garden Trimmings	60	4.5
9	Wood	20	1.5

(continued)

(continued)

S. No.	Component	Moisture %	Ash%
10	Miscellaneous Organics	25	5
11	Glass	2	0
12	Tin cans	3	0
13	Non-ferrous metals	2	0
14	Ferrous metal	3	0
15	Yard waste	60	0.5
	Leaves	30	

Source https://www.researchgate.net/publication/337137470_International_Journal_of_Scientific_Research_and_Reviews_Assessment_of_approximate_chemical_formula_and_energy_content_by_Modified_Dulong_formula_of_municipal_solid_waste_of_Allahabad_city/figures?lo=1

Annexure-II

Typical Density of Municipal Solid Waste (MSW) Components

Component	Density in Kg/m^3
Food waste	290
Paper	85
Cardboard	50
Plastics	65
Yard wastes	105
Wood	240
Tin cans	90

Annexure-III

Typical Energy Content of Municipal Solid Waste (MSW) Components

Component	Energy content KJ/Kg
Food waste	4650
Paper	16,750
Cardboard	16,300
Plastics	32,600
Yard wastes	6500
Wood	18,600
Tin cans	700

Source https://www.chegg.com/homework-help/questions-and-answers/problem-1-energy-content-dry-basis-ash-free-basis-solid-waste-sample-composition-given--as-q55549283#question-transcript

Annexure-IV

Main Types of Agricultural Substrates Utilized by the Biogas Plants Worldwide and Their Properties

[FM: Fresh Matter. NL: normal liter, FM: fresh matter, t: ton.]

Feed-stock	Biogas yield M^3/Ton	Energy value MJ/M^3	Energy value MJ/Ton	Energy value MJ/Kg
Cattle slurry	25	24	600	0.6
Pig slurry	26	23	598	0.598
Laying hen litter	120	25	3000	3
Broiler manure	75	22	1650	1.650
Food waste	46	23	1058	1.058

Source https://slideplayer.com/slide/13726896/

Annexure-V

Chemical properties—ultimate analysis						
Component	Percent by weight (dry basis)					
	Carbon	Hydrogen	Oxygen	Nitrogen	Sulfur	Ash
Organic						
Food wastes	48.0	6.4	37.6	2.6	0.4	5.0
Paper	43.5	6.0	44.0	0.3	0.2	6.0
Cardboard	44.0	5.9	44.6	0.3	0.2	5.0
Plastics	60.0	7.2	22.8	–	–	10.0
Textiles	55.0	6.6	31.2	4.6	0.15	2.5
Rubber	78.0	10.0	–	2.0	–	10.0
Leather	60.0	8.0	11.6	10.0	0.4	10.0
Yard wates	47.8	6.0	38.0	3.4	0.3	4.5
Wood	49.5	6.0	42.7	0.2	0.1	1.5
Inorganic						
Glass	0.5	0.1		< 0.1	–	98.9
Metals	4.5	0.6		< 0.1	–	90.5
Dirt, ash, etc.	26.3	3.0		0.5	–	68.0

Annexure-VI

Load that 1 M^3 of Bio-Gas Can Operate

Load	No. of hours
Electricity generation	1.25 KW
Cooking	4 Persons (each person 0.3 M^3)
2HP engine	1 Hour
60Watts lamp	7 Hours
300Litre refrigerator	3 Hours

Annexure-VII

Electrical Energy Consumption of Various Appliances

Application	Biogas
Cooking meals per person/day	0.3 M^3 (from 0.28 to 0.42 M^3, i.e., 10 to 15 cu ft)
Biogas lamps (averages) Cambodia Ethiopia India	0.049 m^3/h 0.048 m^3/h 0.093 m^3/h
Biogas stove burners (minimum standards) China India Kenya	0.380 m^3/h 0.450 m^3/h 0.500 m^3/h

References

1. https://www.sustainablesanantonio.com/practices-technology/reduce-reuse-recycle/
2. https://www.conserve-energy-future.com/reduce-reuse-recycle.php
3. https://nitsri.ac.in/Department/Chemical%20Engineering/BRTL8.pdf
4. https://sbm.gov.in/sbmReport/Report/Physical/SBM_TargetVsAchievementWithout1314.aspx
5. Abarghaz Y, Mahi M, Werner C, Bendaou N, Fekhaoui M Evaluation of formulas to calculate biogas production under Moroccan conditions. This paper presents different formulas for calculating biogas production and evaluates them for use in the Moroccan context. http://www.ecosan.at/ssp/issue-09-biogas-systems/SSP-09_Okt2011_18-23.pdf
6. Measuring small-scale biogas capacity and production. https://www.irena.org//media/Files/IRENA/Agency/Publication/2016/IRENA_Statistics_Measuring_small-scale_biogas_2016.pdf
7. Biogas calculation tool user's guide Alternative Energy Promotion Center—NRREP Government of Nepal June 2014. https://www.aepc.gov.np/uploads/docs/2018-06-19_Biogas%20calculation%20Tool%20User's%20Guide,%202014.pdf
8. https://www.researchgate.net/publication/337137470_International_Journal_of_Scientific_Research_and_Reviews_Assessment_of_approximate_chemical_formula_and_energy_content_by_Modified_Dulong_formula_of_municipal_solid_waste_of_Allahabad_city/figures?lo=1
9. Sarlahi, NBM, Jha AK (2016) Estimation of biogas potential and sizing of anaerobic digester for malangwa municipality. In: Proceedings of IOE graduate conference, pp 163–169. https://www.researchgate.net/publication/319650145_Estimation_of_Biogas_Potential_and_Sizing_of_Anaerobic_Digester_for_Malangwa_Municipality_Sarlahi_Nepal
10. Hafner SD, Rennuit C (2019) Predicting methane and biogas production with the biogas package. https://cran.r-project.org/web/packages/biogas/vignettes/predBg_function.pdf
11. https://energypedia.info/wiki/Sizing_of_the_Biogas_Plant- Sizing of the Biogas Plant
12. Sasse L (1988) Biogas plants. A Publication of the Deutsches Zentrum für—GATE in: Deutsche Gesellschaft für Technische Zusammenarbeit (GTZ) GmbH. https://www.susana.org/_resources/documents/default/2-1799-biogasplants.pdf

13. https://www.irena.org//media/Files/IRENA/Agency/Publication/2016/IRENA_Statistics_Measuring_small-scale_biogas_2016.pdf
14. https://www.bigrentz.com/blog/smart-waste-management
15. Ioan B, Amellitta L (2012) Mechatronic incinerator for solid non- recyclable waste with energy recovery. ANNALS of the Oradea University. Fascicle of Management and Technological Engineering XI(XXI):NR2
16. https://greensutra.in/waste-segregation-all-you-need-to-know/
17. https://roboticsandautomationnews.com/2019/05/09/5-technologies-that-would-change-waste-management-industry/22289/
18. https://www.ciwm.co.uk/ciwm/knowledge/the-r1-energy-efficiency-formula.asp
19. Abdel-Magid IM, Problem solving in solid waste engineering
20. Anschutz J, Ijgosse J, Scheinberg A (2004) Putting integrated sustainable waste management into practice. Using the ISWM assessment methodology. Waste, Gouda, The Netherlands
21. de Bertoldi M (1996) Science of composting. Springer
22. CEHA, Solid waste management in some countries of the Eastern Mediterranean region, WHO, Eastern Mediterranean Regional Office, Regional Centre for Environmental Health Activities, Amman, Jordan, CEHA Document No. , Special studies, ss-4, 1995
23. HS Peavy DR Rowe G Tchobanglous 1985 Environmental engineering McGraw-Hill Book Co. New York
24. Perry RH, Green DW, Maloney JO, Edi Perry's (2007) Chemical engineers' handbook (8th ed). McGraw-Hill Professional
25. Worrell WA, Vesilind PA (2012) Solid waste engineering. CL-Engineering Pub

Chapter 4
Grid Integration of Renewable Energy System (GIRES)

4.1 Introduction

The proliferation of renewable energy systems into the power grid requires upgraded systems and methods to effectively and efficiently balance the supply from the grid and the renewables to the various loads. Integration of renewable electricity generation plants to distribution grids requires the analysis [1] of several parameter factors [IRENA-2021] which impact the grid's operation. A major criterion for GIRES is the impact on the grid voltage during normal operations.

The various renewable energy sources that can be integrated with the grid are, to a large extent solar, wind and waste-to-energy plants. To a lesser extent we have ocean, tidal, wave and geo-thermal energy plants.

4.1.1 Advantages of Grid Integration of Renewable Energy System (GIRES)

Before we delve into the issues of goals, planning and maintenance of GIRES, we shall see some of the critical advantages of some renewable energy source techniques. The Most convenient and economical renewable sources, which can be commissioned by an individual [2] are solar and wind. Other sources as Ocean, Tidal and geothermal can be commissioned by Governments. Hence, we shall limit our present studies to solar and wind resources only.

4.1.1.1 Solar Roof Top Photo-Voltaic (PV) Systems—Advantages

Electricity generated by Solar PV systems installed on rooftops of a building-could be residential, commercial, institutional and industrial buildings:

N. Manohar et al., *Machine Learning Applications in Renewable Energy*, Green Energy and Technology, https://doi.org/10.1007/978-981-97-9939-8_4

- Reduce transmission and distribution losses.
- Have low gestation time.
- Do not require additional land.
- Additional Electricity generated i.e. if generation is greater than consumption, can be—fed into the grid at regulated feed-in tariffs.
- Improve grid voltage levels.
- Increases employment opportunities.
- Reduces power bill by supplying surplus electricity to local electricity supplier
- Installation is easy.
- Cost of system is optimum.
- Easy and minimal maintenance.

4.1.1.2 Wind Turbines

- **Wind is almost everywhere**
- **Wind is consistent in the medium and long-term**
- **Excellent conversion efficiency**
- **Wind power occupies very little land**
- **The environmental impact is minimal**
- **A green source that is truly economical**
- **Maintenance is simple and low operating costs**
  - Mini-wind power and micro-wind power for domestic purpose.

4.1.1.3 Wave Energy

- Wave Energy is a Renewable Source.
- It is Environment Friendly, indicating that it causes less atmospheric pollution.
- It is Abundant and Widely Available in almost any location.
- The advancements in technology has made available Various Ways To Harness wave energy.
- It is predictable with the assistance of meteorological department.
- It causes no Damage to Land.
- Wave energy is a reliable source.

Summarizing the advantages of grid integration of renewable energy technologies, we have:

- Grid integration maximizes cost-benefits of power generation, transmission and distribution by optimizing the utilization of variable renewable energy sources.
- Grid integration enhances power system stability and reliability.
- Reduction in carbon emissions, contributing to environment cleanliness.
- Reduces power system losses.

4.1.2 *Goals of Grid Integration of Renewable Energy System (GIRES)*

Advancement of system design, planning, and operation of the electric grid for effective integration of renewable has the goals as indicated below:

- Reduce carbon emissions and other air pollutants through increased use of renewable energy.
- Increased use of assets, through integration of distributed systems and customer loads to reduce peak load and thus lower the costs of power.
- Support standardization of standards for renewable energy and energy efficiency.
- Enhance reliability, security, and resiliency of microgrid applications even in critical situations and highly constrained areas of the electric grid.
- Support reductions in oil use by enabling plug-in electric vehicle (PHEV) operations with the grid.

4.1.3 *Grid Integration Planning*

The effective planning methodology for grid integration needs to consider the system requirement, technical, operational and financial aspects [3]. These aspects are covered under the three areas:

1. Capacity Expansion—Study on scope of expansion
2. Power Flow Studies—Technical studies on electrical operational parameters.
3. Technical Specification computation
4. Cost Analysis.

1. **Capacity Expansion—Study on Scope of Expansion**

Studies are conducted to identify the gap between supply and demand, and the scope for capacity expansion. The study indicates where, when, how much, and what types of infrastructure (generation and/or transmission) is required to achieve the target of minimizing the gap between supply and demand, in terms of energy requirement quantity and cost, taking into consideration factors such as new policies, technological advancement, fuel prices, and demand projections.

2. **Power Flow Studies—Technical Studies on Electrical Operational Parameters**

Power flow studies or load flow studies are performed to estimate the state of various electrical parameters such as voltage, current, and power in a power grid [4]. These estimates allow for efficient and reliable management of a given grid under both stable and uncertain conditions to keep the grid state stable. The rapid advancement of power system and computing equipment hardware technology requires the use of fast data capture, calculation of parameters, analysis and sending corrective action signals to various decision-making [5], control and execution devices. With the advent of artificial intelligence technology, this has become possible.

3. **Technical Specification Computation**

Renewable energy systems are usually set up for a specific organization. Hence, first the electrical energy requirements of the organization, called the Customer Requirement Specification (CRS) is to be determined. Having got the Technical CRS, the budget requirement or the Financial requirement is required to be estimated based on standards and market availability of the various components.

4. **Cost Analysis**

Having identified the need for capacity expansion and conducted power-flow studies, the next step is determining the technical details of the renewable energy system to meet the capacity expansion and thereby compute the cost. The budget requirement or the Financial requirement is estimated based on standards and market availability of the various components.

4.1.4 Challenges of GIRES

Challenges of GIRES are:

4.1.4.1 Instantaneous to Seasonal Unavailability of Renewable Resources—Intermittency

As we all know, renewable energy is available 24 h a day[6], but its availability is not the same every day at a particular time. Furthermore, although solar energy is abundant in many locations, cloud cover, rainfall, dust storms, etc. can cause significant changes in solar output. Therefore, necessary measures should be taken to handle this intermittency, such as storing the renewable energy when it is higher than the load requirement and supplying power from the grid or storage if the renewable energy is lower than the load requirement. Mathematically the situation is expressed in Eqs. 4.1 and 4.2.

$$P_{RES} > P_D;\ P_D \text{ is catered through RES and } B_C \tag{4.1}$$

$$P_{RES} < P_D;\ P_D \text{ is catered through grid or } B_D \tag{4.2}$$

where

P_{RES} Power from renewable energy sources
P_D Power demand (Load)
B_C Battery or storage system charging

B_D Battery discharge/load supplied by battery.

4.1.4.2 Framing of Standards

The integration of renewable energy into the grid requires large-scale transformation of the existing network [7] and the development of the following standards: (a) Allowing bidirectional energy flow, Fig. 4.1 That is, the installation of distributed generation from top to bottom (from power plants to consumers) and from bottom to top (end users providing electricity supply) with the aim of ensuring grid stability establish an efficient electricity-demand and grid management mechanism aimed at [8] reducing peak loads, improving grid flexibility, responsiveness, and security of supply (b) improve the interconnection of grids at the regional, national and international level, aimed at increasing grid balancing capabilities, reliability, and stability (c) introduce technologies and procedures to ensure proper grid operation stability and control (e.g. frequency, voltage, power balance) in the case of large shares of renewable energy, and (d) When electricity supply exceeds demand, energy storage capacity is introduced to store electricity from renewable energy sources, as shown in Fig. 4.1 and satisfying Formulas 4.1 and 4.2.

IEEE 1547 is the Standard for Interconnection and Interoperability of Distributed Energy Resources with Associated Electric Power Systems Interfaces. Interfaces as defined in IEEE 2030 is "a logical interconnection from one entity to another that supports one or more data flows implemented with one or more data links." IEEE 1547 provides requirements relevant to the interconnection and interoperability performance, operation, and testing, and, safety, maintenance and security considerations.

The standards for safe electricity use in India are:

1. IS732: Code and Practice for Electrical Wiring Installation.

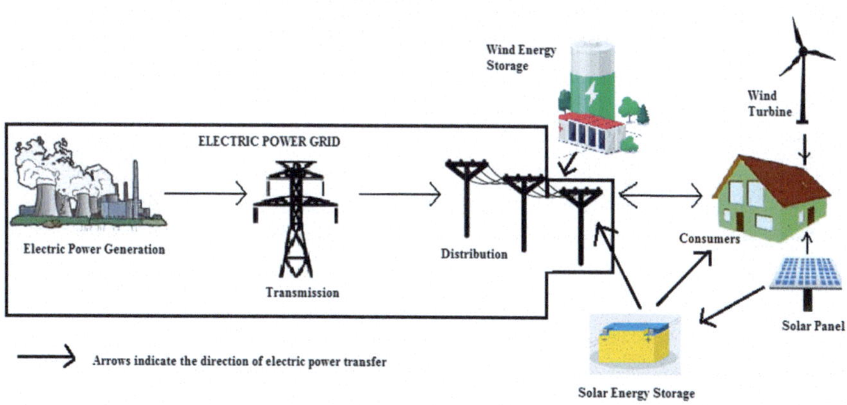

Fig. 4.1 Hybrid power system with bi-directional flow of electric power

Table 4.1 UL standards used for the BOS components

Description	UL standards
PV modules testing and certification	UL 1703 IEC/EN 61215, IEC/EN 61646 IEC 61730/UL 790 (Fire Test) UL-SU 8703 (CPV)
Mounting system	UL-SU 1703–A
Tracking system	UL-SU 9703
Combiner box	UL 1741
Inverter	UL 1741 IEC 62109
Batteries	UL 1989, UL 2054 UL-SU 2580, UL-SU 1973

2. IS12360: Voltage Bands for Electrical Installations Including Preferred Voltages and Frequency.
3. IEEE519-2014: IEEE Recommended Practice and Requirements for Harmonic Control in Electric Power Systems.

To ensure the consistency, safety and quality of Solar PV system components and to increase consumer confidence in system performance IEEE, Underwriters Laboratories (UL), International Electrotechnical Commission (IEC), AM0 Spectrum (ASTM) are working on standards and performance criteria for PV systems.

Balance of system (BOS) refers to all equipment required to operate and integrate PV panels into a grid-tied or off-grid system. The BOS includes all of the components of a PV system except for the solar panels themselves. A similar term to Balance of System (BOS) is "Balance of plant (BOP)" which is generally used in the context of power engineering and applies to all the supporting components and systems of the power plant which are needed to produce the energy. These may include suitable transformers, inverters, cabling, switching and control equipment, protection equipment, power conditioners, support structures, etc., depending on the type of plan. The UL standards used for the corresponding BOS components of the Solar PV system are as shown in the above Table 4.1.

4.1.4.3 Building Resilience Against Threats, Such as Natural Disasters and Cyber Threats

The digitalization of the energy industry increases its vulnerability to cyber and terrorist attacks, one of which occurred in 2022. Lithuania's state-owned energy group Ignitis, Ukraine's state nuclear power company Energoatom, and Greece's largest natural gas supplier DESFA were among the energy companies hit by cyber attacks. And there was an increase in physical attacks on power substations in the US.

Table 4.2 Categories of PQ problems [9]

Power freq disturbance	Electro magnetic interferences	Power system transient	Power system harmonics	Electrostatic discharge	Power factor
i. Low freq phenomena ii. Voltage sag/swell	i. High freq phenomena ii. Interaction between electric and magnetic field	i. Fast, short duration event ii. Produce distortion like notch, impulse	i. Low frequency phenomena ii. Produce waveform distortion	i. Current flow with different potentials caused by direct current or induced electro-static field	i. Low power factor causes equipment damage

4.1.4.4 Maintaining Required Voltage Levels Within Regulated Limits

Voltage levels should be maintained within the limits prescribed by the respective State Electricity Commission specifications. Maintaining voltage and frequency levels will ensure power system stability. Power system stability is defined as "the property of a power system that enables it to remain in a state of operating equilibrium under normal operating conditions and to regain an acceptable state of equilibrium after being subjected to a disturbance." Disturbances can be small or large.

India follows the IS12360 standard that requires low voltage single phase supply to be delivered at 230 V, with the minimum and maximum value ranging from 207 to 253 V. In addition to this the IEEE519-2014 requires the power quality measured in terms of Total Harmonic Distortion (THD) of Voltage should be less than 8%. Most cases these voltage quality parameters are not met which leads to unwanted failures and loss of assets. The result of the failure to meet voltage standards are burning of fridge, televisions, air-conditioners, motors and any electrical appliance, causing severe damage as the entire premises catching fire. The loss is not only of products and equipments but also of personnel if any in the premises during that time.

Voltage changes are identified by complex load-flow calculations. The voltage increase depends on the network short-circuit power and the impedance angle at the connection point. It also relates to equipment characteristics (i.e. transformers, distribution lines) and is affected by the phase angle (ϕ) of the plant, which can be adjusted by reactive power control.

4.1.4.5 Coordinating—Protection System Devices

The protection system devices as relays and circuit breakers, coordination is very important, to meet the objectives as:

- Have minimum outage.

- Fast restoration of fault devices of power system.
- Minimize the extent of damage due to any fault.

The existing protection schemes are not adaptive due to the different ways of the current flow due to different operations. The entry of renewable resources into the power grid will change the system coordination and normal protection plans will become completely ineffective. Applied protection schemes based on independent decisions will not be able to handle the natural complexity of the power grid. Innovative protection schemes based on data sharing are needed to protect smart grids. The present day technology as microprocessor based protection devices has increased the effectiveness of coordination.

4.1.4.6 Overloading of Existing Transmission Lines

A transmission line has a specified current carrying capacity, If the passing current exceeds the limit, thermal loads will build up, leading to damage. The unusual situation may occur when (i) any of producers generate too much power without warning, and the entire or part of the power system would shut down. (ii) some heavy loads go off the grid without prior intimation to concerned authorities.

Hence, prior planning is very important to prevent overloading of transmission lines.

4.1.5 *Measures to Mitigate Challenges of GIRES*

- Flexible generation,
- Ancillary services for supply-balancing,
- Standardization of technical requirements—grid specifications, connexion standards, real time monitoring etc.
- Demand side management—demand response and storage for load balancing,
- Forecasting of renewable generation and forecasting of demand,
- PMUs/WAMS on pooling stations and interconnection with centralized control centre for real time information, monitoring, and control,
- Policy and regulatory advocacy for electricity balancing markets and pricing mechanisms
- Renewable Energy Management Centers (REMC)—A dynamic load flow management system (which seems to be the best option) found in a unified power flow controller that can be fast-reacting. The solution should keep the power line within the n – 1 emergency standard balance range by managing series and shunt compensation, thus keeping power flowing at an optimal level.
- Usage of energy storage technologies,
- Implementation of Smart grids,

- Implementation of fast response and rigid interfacing and protection devices and control strategies, by way of adoption Artificial Neural Networks, Internet-of-Things and modern digital electronic devices.
- Preventive analysis of future grid violations—utilizing the data and analytics available on the grid, preventative measures can be derived and put in place to avoid breaches.
- Install a large number of reactive power compensation devices and build high-voltage direct current transmission lines from the power generation centre to the load centre.

4.1.6 RES Grid Integration—Quality Issues

Renewable energy source grid integration need to be studied from technical, operational, maintenance and financial point of view. In this section, technical aspect is presented. Depending up on the generation capacity of the renewable energy source, grid integration of the systems shall be done at the transmission level (large capacity) or at the distribution level (small capacity). Currently, most renewable energy systems are connected at the low voltage distribution level as shown in the Fig. 4.2. The technical issues depend on the type of renewable energy source also. However, most of the issues are common to all types of sources.

The technical issues that shall be addressed while integrating RE resources are:

1. Point of common coupling (PCC)
2. Voltage variations and power quality

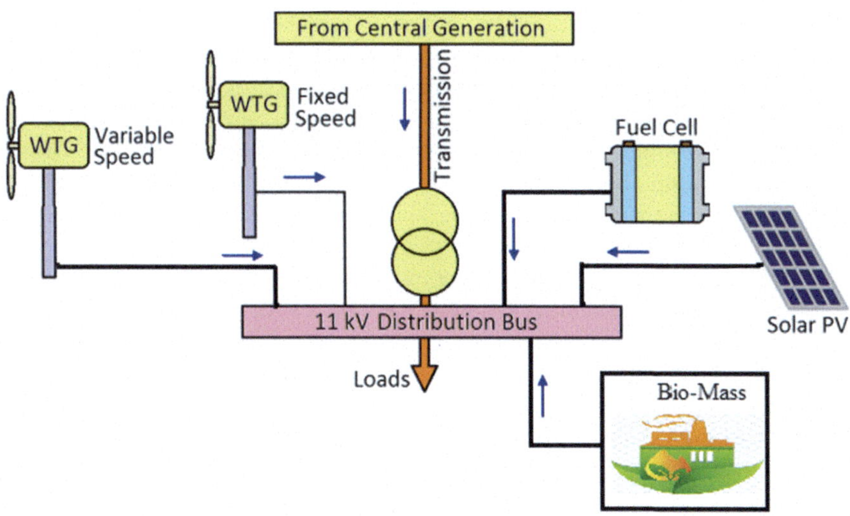

Fig. 4.2 Grid connected distributed renewable resources

3. Voltage ride-through capability
4. Reactive power compensation capability
5. Frequency regulation capability
6. Protection issues
7. Voltage variations and power quality.

- Voltage fluctuations
- Electromagnetic interference
- Power system transients and harmonics
- Voltage stability problems
- Weak grids
- Switching losses
- Synchronisation
- Grid congestion.

4.1.7 Grid Integration Terminology [10]

Balancing area: the collection of generation, transmission, and loads within the metered boundaries of the responsible entity (i.e., the balancing authority) that maintains balance between electricity supply and demand within this boundary.

Capacity value: the contribution of a power plant to reliably meet demand, measured either in terms of physical capacity (kW, MW, or GW) or as a fraction of the power plant's nameplate capacity (%).

Demand response: voluntary (and compensated) load reduction used as a system reliability resource.

Flexibility: the ability of a power system to respond to changes in electricity demand and supply.

Grid integration of renewable energy: the practice of power system planning, interconnection, and operation that enables efficient and cost-effective use of renewable energy while maintaining the stability and reliability of electricity delivery.

Grid integration study: an analysis of a set of scenarios and sensitivities that seeks to inform the stakeholders on the ability and needs of a power system to accommodate significant VRE.

Storage: technologies capable of storing electricity generated at one time and for use at a later time. **Variable renewable energy (VRE)**: electricity generation technologies whose primary energy source varies over time and cannot easily be stored. VRE sources include solar, wind, ocean, and some hydropower generation technologies.

Uncertainty: the inability to perfectly predict electricity demand and/or generator output.

Variability: the changes in power demand and/or the output of a generator due to underlying fluctuations in resource or load.

References

1. https://www.irena.org/-/media/Files/IRENA/Agency/Publication/2015/IRENA-ETSAP_Tech_Brief_Power_Grid_Integration_2015.pdf
2. https://www.adb.org/sites/default/files/publication/183785/sdwp-043.pdf
3. https://www.nrel.gov/docs/fy16osti/66504.pdf
4. https://www.greeningthegrid.org/Grid-Integration-Toolkit/grid-integration-guidebook/types-of-grid-integration-analyses
5. https://www.igi-global.com/chapter/load-flow-analysis-in-smart-grids/184022
6. https://www.researchgate.net/publication/228303570_Intermittency_and_the_Value_of_Renewable_Energy
7. https://www.nrel.gov/grid/standards-codes.html
8. https://www.researchgate.net/publication/322750538_Grid_Integration_of_Renewables_Challenges_and_Solutions
9. https://www.icrepq.com/icrepq'10/505-Khadem.pdf
10. https://www.nrel.gov/docs/fy15osti/63033.pdf

Chapter 5
Artificial Intelligence and IoT in Renewable Energy

5.1 Introduction

The dynamic grid of the present day with integration of renewable energy sources of the government and general public, development of technologies as SCADA are capturing large data that need to be analyzed to be able to give corrective action signals to maintain power system stability. There is therefore need to determine the optimum corrective action that need to be taken within a few micro-seconds.

As several megawatts feed into the grid from variable renewable energy sources, determining the capacity levels is of prime importance for stable and efficient grid operation. Renewable taking up a greater share of the grid, there is a loss of base load generation from sources such as coal, which provide grid inertia through heavy rotating equipment such as steam and gas turbines. Without grid inertia, power networks will be unstable and susceptible to blackouts. Supported by emerging technologies, such as the internet of things (IoT), sensors, big data and distributed technology, the modern day sensors are able to capture enormous amount of real-time data, which can be processed and analyzed only with high speed software applications as artificial intelligence programs. The powerful prediction capabilities [1] of AI will lead to improved demand forecasting and asset management. The automation capability of AI can lead to operational excellence in many crucial areas.

Some major companies that are leading in technological development and implementation of AI in the energy sector include IBM Corporation, Intel Corp, Cisco Systems Inc., Mitsubishi Hitachi Power Systems Ltd, Microsoft Corporation.

5.1.1 AI Techniques

The various AI Techniques [2] that are applicable for renewable energy system are:

1. Feed-forward back propagation neural network (BPNN)

N. Manohar et al., *Machine Learning Applications in Renewable Energy*, Green Energy and Technology, https://doi.org/10.1007/978-981-97-9939-8_5

2. Adaptive Neuro-Fuzzy Inference System (ANFIS)
3. Genetic Algorithm (GA)
4. Artificial Neural Network (ANN)
5. Radial basis function neural networks (RBFNN)
6. Bacterial foraging Optimization (BFO)
7. Particle Swarm optimization (PSO)
8. ARIMA (Auto Regressive Integrated Moving Average) time series algorithm using continuous data.
9. Machine learning algorithm using regression tree model
10. Sugeno type fuzzy system
11. Q-learning algorithm
12. Deep Learning.

5.1.2 AI Application Areas

AI application areas [3, 4] in RES can be categorized into three, based on the supplier, process and customer perspective as:

i. RE Generation side
ii. Power system operation process
iii. Consumer side.

1. **Demand forecasts**: Forecasting [5] application like short term and long term load forecasting, electricity market forecasting, solar power forecasting, wind power forecasting, enable forecasting the demand and thereby adjust the generation from various sources to reduce lasses and enhance power system reliability and stability.

Tata Power uses AI to predict solar energy production of its solar power plant. This helps the company better manage its grid and ensure it has enough power to meet demand. To forecast wind energy generation and solar energy generation Google is developing AI algorithms.

2. **Asset maintenance**

AI-powered systems can analyze market data, energy prices, and demand forecasts to optimize energy trading strategies. This allows renewable energy operators to make informed decisions and maximize revenue while contributing to grid stability. Power Grid Corporation of India (PGCIL): is using artificial intelligence to improve the management of the power grid. This helps the company better integrate renewable energy into the grid and instantly balance supply and demand.

3. **Help customers control energy costs**

MIT researchers are using AI to develop new solar cell materials that are more efficient and less expensive to produce.

4. **Tuning of the significant parameters'**

Several companies have improved the hybrid renewable energy system (HRES) parameter performance applying AI techniques [6]. The company, renewable energy source and the parameter enhance in presented in Table 5.1. The table provides only a few representative data.

5. **Energy management**

Energy Management involves the balancing of generation and load, such that in an ideal situation (i.e. a state of the power system where there are no losses):

$$\text{Power Generated} = \text{Load Demand (i.e. Consumer requirement)} \tag{5.1}$$

where, Power Generated is the sum total of the power generated by all renewable sources.

However, we know that no system is practically ideal. Hence, in a hybrid renewable energy system (HRES), the condition would be:

$$\text{Power Generated} = \text{Load Demand (i.e. Consumer requirement)} + \text{Losses}$$

Considering a HRES with grid supply, solar and wind as in Fig. 5.1, if the total power generation from all sources is more than the demand, then the excess power than required from solar and wind would be stored in batteries.

AI is being used by Tesla [8] to develop solar energy storage management system. These systems control the storage of solar energy generated during the day and used at night or when the sun is not shining.

Table 5.1 Company, renewable energy source and the parameter enhanced with AI

Company	RES	Parameter improved with AI
Siemens gamesa renewable energy	Wind turbines	Improve efficiency of wind turbine and thereby increase lifespan
DeepMind	Solar and wind	Developing algorithms to optimize the performance

Fig. 5.1 Wind energy, solar energy and storage system

Autogrid is using AI to develop software to help grid operators manage the grid with a high share of renewable energy.

6. **Increase efficiency**: The efficiency of the renewable energy [7] system can be substantially improved by applying AI Techniques, to adjust the output and input by continuous monitoring the weather conditions.

ReNew Power uses artificial intelligence to increase the efficiency of wind turbines. This helps the company produce more electricity from wind farms.

7. **Wind energy: planning**—Wind turbine positioning, reactive power optimization, network feeder routing, and capacitor placement.
8. **Energy storage management** decisions can be done effectively [9] by considering forecast demand, renewable energy generation, prices and network congestion, among other variables.

Battery storage is a technology that enables power system operators and utilities manage and control energy storage for later use. A battery energy storage system (BESS) is an electrochemical device that charges (or collects energy) from the grid or a renewable energy source and then discharges that energy at a later time to provide electricity to the consumers.

- Tesla uses AI to develop solar energy storage systems that can be used to store solar energy generated during the day and used at night or when solar energy cannot be obtained, due to insufficient solar radiation levels. Figure 5.1 shows how the excess wind and solar energy generated beyond load demand requirement is stored in batteries by Tesla.

9. **Loss minimization**: The losses in a renewable energy system can be minimized using artificial intelligence techniques.
10. **Grid management**

AI algorithms analyze data on weather patterns, energy demand, and grid operation conditions and determine the optimal allocation of renewable energy resources. This helps maximize energy production, minimizing costs and also meet the requirements of the users without any energy supply interruption.

- Autogrid [10] is developing AI software to help grid operators manage the grid with a high share of renewable energy.
- Power Grid Corporation of India Limited (PGCIL) [11] is using AI to improve the power grid management in India.
- **Maintenance scheduling and preventive maintenance**

Preventive Maintenance: Preventive Maintenance is essential to avoid any component or system failure. Predictions on power system health can be done by collecting data from different sensors and performs in-depth analysis to predict the likely occurrence of failure and thereby take necessary action in advance. AI modeling enable optimal maintenance operations and asset protection, preventing costly and potentially dangerous downtime and asset failures.

The companies using AI Techniques for preventive maintenance are:

- O&M for solar power plants. This helps reduce time and costs.
- General Electric [12] has developed a software named Predix, to build a sustainable and reliable grid of renewable energy.
- Vestas is using for wind turbines.
- SolarEdge for solar power plants.
- **Quality analysis**—Renewable energy sources are intermittent by nature, which effects grid stability. AI algorithms [13] can predict fluctuations in the quality and quantity of the input resource which is sun, wind and tidal intensity and thereby adjust dynamically grid parameters to maintain stability, ensuring a smooth integration of renewable sources, reduce Harmonics and control power flow in the entire power system from sources to users.

One of the major aspect to be addressed in a power system with RES is the generation of harmonics in inverters, which are an important device as they convert the Direct Current (DC) power output of the renewable energy source to Alternating Current (AC) power, as most loads operate on AC.

5.1.3 Advantages of AI Combined with IOT in RES

AI combined with IoT can help:

- **Optimize energy distribution and storage**: IoT can help with remote controlling of systems, consumption tracking, data collection, and analysis and AI can help with decision making about where to supply energy, which generating capacity to activate or turn off the capacity of energy to store, and many other decisions.

Through Artificial Intelligence, energy storage can be optimized. Storing renewable energy is challenging since the production of this energy is chaotic and unpredictable. Combining renewable energy with AI-enable storage can enhance energy storage management thus boosting power system efficiency while reducing power losses.

- **Auto-monitoring and marketing power plant**

Most of the data processing and forecast is done in the Virtual Power Plant. Artificial Intelligence algorithms help to generate accurate forecasts while coordinating various systems in the Virtual Power Plant. The technology helps to determine which plant generates or uses how much power and when- based on data from power trading stations, live-feed-in data, weather forecasts, historical data, etc.

Some AI algorithms are sufficiently intelligent [14] that they can trade electricity without human intervention through the process known as automated trading, algorithmic trading, or algo- trading. AI is also helping with auto-monitoring and analysis of trade on the power market.

- Improve the accuracy and speed of power quality analysis.
- Detect and diagnose power quality issues in real-time, allowing for faster and more accurate resolution.
- Predict potential power quality issues before they occur, allowing for proactive management and prevention of power quality issues.
- Improve the efficiency of power quality management.
- By leveraging AI-based optimization algorithms, utilities can reduce energy consumption and costs while improving power quality.
- Areas of the grid where energy is being wasted and make adjustments to reduce
- They have a high level of fault tolerance.
- ANNs are quick and dependable.
- They have the ability to learn and adapt to new information, as well as the ability to generalize.

5.1.4 Artificial Intelligence's Challenges in the Renewable Energy Sector

The challenges of AI with IOT challenges are:

1. Requirement Specification

Many energy companies lack enough technical background knowledge that could help them understand how AI with IOT can improve the performance of their operations and benefit their companies. They are therefore not able to specify their actual requirement.

Many companies are, therefore, still operating with time-proven methods and tools instead of risking adopting new technologies.

2. Insufficient Experienced Personnel

AI with IOT technology is still in a budding stage. That is why only a few mastered professionals are available. Even though there are experts [15] with ample theoretical knowledge of the subject, only a few professionals are able to design powerful AI-IOT enabled systems with real practical value to the industry.

3. Lack of Finance

Energy companies are struggling with financial pressure for adopting innovative smart technology, even though it's the best option.

Besides, finding a reliable software service provider, creating and customizing software, adjusting, managing, and monitoring it takes time and resources.

To reap the benefits of incorporating AI and other technologies into their strategies, energy companies and businesses need to allocate a good amount of budget and accept the risks that come with switching from their outdated to modern energy systems.

5.1.5 *Future Impact of AI on Renewable Energy Systems*

Several standard organizations have analyzed and predicted the impact of AI on Renewable Energy Systems of the future. Some of the predictions are:

- The International Renewable Energy Agency (IREA [16] predicts an energy savings of up to 10% by 2030).
- A study conducted by the National Renewable Energy Laboratory (NREL) shows that solar energy forecasting accuracy would increase by 10–20%.
- According to a report by Wood Mackenzie, wind farms production would be up to 20%.

5.1.6 *Case Studies on Implementation of AI in RES*

5.1.6.1 Power System Stability

Several companies have developed AI based devices and software algorithms to monitor and control the operation of hybrid power system (HPS), to maintain system stability. Knowledge of power system stability is required for integrating of AI Technique into the power grid.

Power system stability is the study of the dynamics of the power system under disturbances. Power system stability is the ability of the power system to return to normal or stable operation after having been subjected to some form of disturbances. The normal state of the power system can be described by the conditions of the power system, as in Table 5.2. The table presents two compulsory parameters, but however based on the size of the power system and operating scenario as occurrence of type of fault, the conditions need to be computed from the measured parameters as voltage and current values and data obtained from various sensors and necessary corrective action shall be taken.

Table 5.2 Conditions describing normal state of the power system

Description of condition	Mathematical expression
Frequency shall be with the specified limits	$f_{min} < f < f_{max}$ (5.2) where f is the measured frequency of the power in Hertz. f_{min} and f_{max} are the minimum and maximum limits of the frequency as per the standards. The standard [17] varies from one country to another. Example in India frequency specification f_{min} is 49.95 Hz and f_{max} is 50.05 Hz
Grid voltage shall be with the specified limits	$V_{min} < V < V_{max}$ (5.3) Where V_{min}, V, V_{max} are the minimum, nominal and maximum limits of voltage in Volts, as specified by standards

Figure 5.2 depicts the flow chart on implementation of AI Technique algorithm for power system stability. The steps of operation are:

1. The output of the sensors [18] is sent to the computational unit of the power system. Figure 5.3 depicts the various types of sensors and their location in a power system. Apart from these sensors of a power grid, there are various other sensors relevant to each type of a renewable energy source, integrated with the grid. Some of the sensors are solar sensor, wind speed sensor, temperature-humidity sensor.
2. The computational unit (CU) may be a Central Processing Unit, microprocessor or microcontroller. The CU computes the status of operation of the power system.
3. The AI Technique algorithms are then applied to the results of the computation to determine the decisive or corrective action.

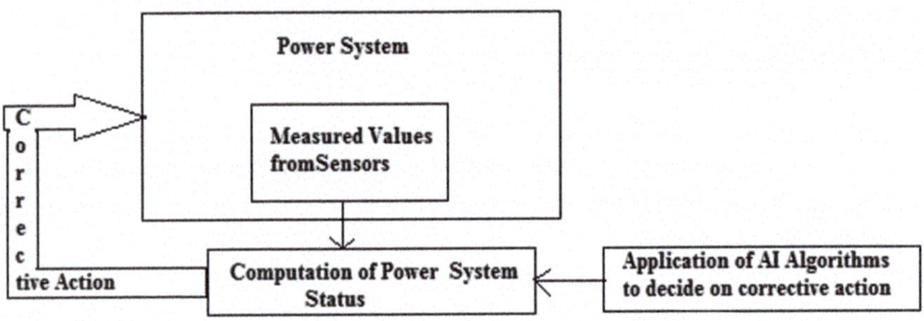

Fig. 5.2 Flow chart of AI algorithm implementation

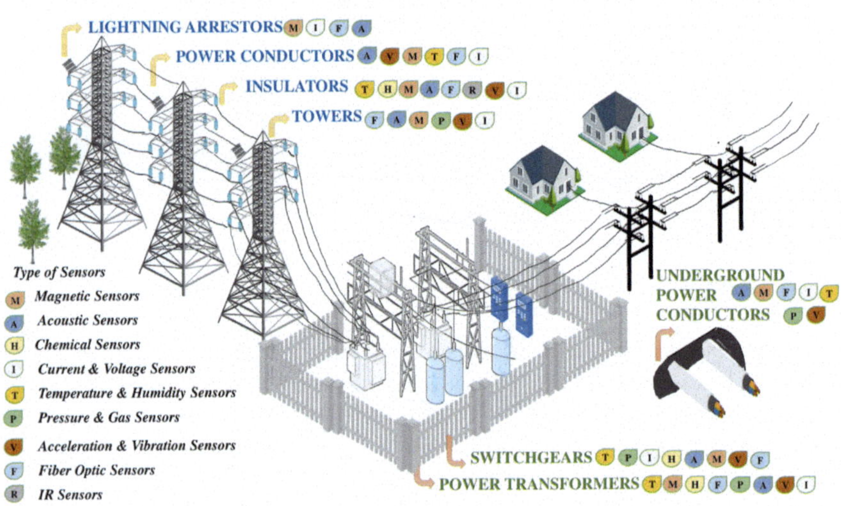

Fig. 5.3 Sensors and their location in power system

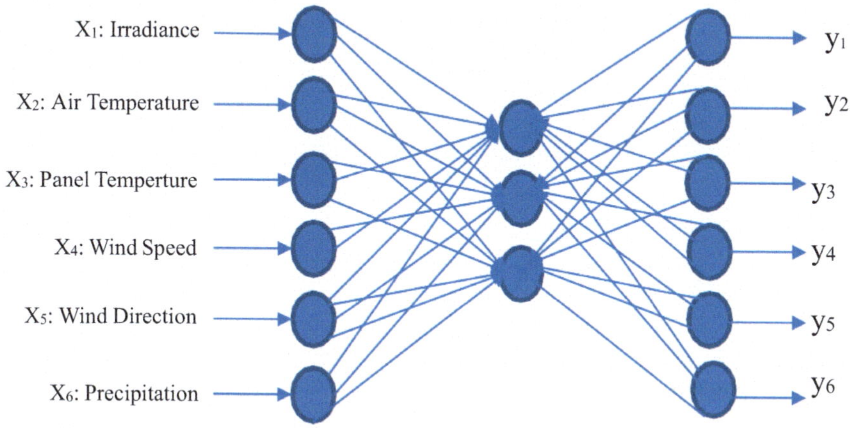

Fig. 5.4 The architecture of multi-layer perceptron NN L

4. The corrective action signal is then forwarded to the corrective device as circuit breaker on the power system or solar energy tracker for the device to operate.

5.1.6.2 Review of Implementation of AI Techniques in Hybrid Power System

Case Study-I

The goal of power system protection is to clear the fault [19] and restore the power system to stable state in the shortest time. The stages of fault management Fig. 5.4 are:

i. Fault detection
ii. Fault state computation and analysis
iii. Classification and identification of fault
iv. Initiate rectification of fault and restoring of power system.

In order to ensure stability of the power system and continuity of service, the faulty area of the circuit should be isolated from the healthy area without time delay. Shuma Adhikar et al.

Case Study—II—Energy Forecasting

Widodo et al. [20] have performed energy forecasting studies of a hybrid-micro grid power system having PV and wind RES, using ML techniques of ANN. Real time data at various locations and various times of the year was collected for the studies.

The forecasting process can be modeled as follows: historical weather time series data in the form (×1, ×2, ..., xn), where n is the number of weather parameters. To predict the solar PV power in the day-ahead category, the first step is to determine the mapping function between historical weather data and solar PV power prediction namely:

$$\text{Output y} = \text{Function of inputs}(\ x1, x2, x3, x4, x5, x6) \tag{5.4}$$

where

Inputs x_1	Irradiance (kW/m^2)
x_2	Air Temperature (°C)
x_3	Panel Temperature (°C)
x_4	Wind Speed (m/s)
x_5	Wind Direction (°)
x_6	Precipitation.

A Multi Layer Perceptron (MLP) Artficial Neural Network which consists of multiple fully connected layers of neurons, has been developed for the prediction analysis. This architecture was then converted into a Deep Neural Network (DNN) architecture with Long Short-Term Memory (LSTM) which was developed from the basic architecture of the Recurrent Neural Network (RNN) as shown in Fig. 5.4.

Several hidden layers were added in this architecture including LSTM memory units. RNN uses temporal information as input data and can make repeated connections between neurons. LSTMs have memory cells in their neurons that have the ability to store information. The information that enters and leaves the neuron's memory cell is controlled by three gates, namely the input gate, the output gate and the forget gate. Each of these gates gets the same input from the input neuron and also has an activation function. The activation function used in this forecasting process can be tanh activaton functions and Rectified Linear Unit (ReLU) activation functions.

The Voltage-Current (V-I) curve data and the meteorological environmental data at the PV Module was collected for the analysis. The dataset was measured at three locations with different climates during one year. Evaluation and verification effectiveness of the proposed Machine Learning algorithm showed good prediction results.

References

1. https://imaginovation.net/blog/artificial-intelligence-in-renewable-energy/
2. https://ieeexplore.ieee.org/document/659867
3. Dellosa JT, Palconit EC (2021) Artificial intelligence (AI) in renewable energy systems: a condensed review of its applications and techniques, 978-1-6654-3613-7/21/$31.00 c2021 IEEE
4. Pandey U, Pathak A, Kumar A, Mondal S (2023) Applications of artificial intelligence in power system operation, control and planning: a review. Clean Energy 7(6):1199–1218.https://doi.org/10.1093/ce/zkad061, Published: 17 November 2023
5. nrel.gov/docs/fy20osti/77593.pdf

6. https://www.solarfeeds.com/mag/advantages-and-disadvantages-of-artificial-intelligence-in-clean-energy/#How_AI_is_transforming_the_energy_industry
7. https://energy5.com/the-role-of-artificial-intelligence-in-optimizing-renewable-energy-systems
8. https://www.cleanenergyreviews.info/blog/powerwall-vs-lg-chem-vs-sonnen-vs-byd
9. https://ampowr.com/energy-storage-systems-wind-turbines/
10. https://www.auto-grid.com/
11. https://www.powergrid.in/en/technology-development-rd
12. https://www.ge.com/news/press-releases/new-ge-predix-software-power-producers-and-utilities-breaks-down-barriers-between
13. https://www.siemens.com/global/en/products/energy/grid-software/pti-consulting/power-system-consulting/power-quality/power-quality-analytics.html
14. https://8msolar.com/integrating-renewable-energy-with-iot-solar-panels-and-smart-homes/
15. https://taraenergy.com/blog/ai-in-energy-advantages-challenges-and-innovations/
16. https://www.irena.org/Digital-Report/Tripling-renewable-power-and-doubling-energy-efficiency-by-2030
17. https://www.generatorsource.com/Voltages_and_Hz_by_Country.aspx
18. Namratha Manohar J, Sensing layer—layered architecture for efficient operation of intelligent power system. In: Emerging IT/ICT and AI technologies affecting society, lecture notes in networks and systems 478. https://doi.org/10.1007/978-981-19-2940-3_1
19. Shuma Adhikar et al, in their study of "Fuzzy logic based on-line fault detection and classification in transmission line", have presented fuzzy logic based online fault detection and classification of transmission line. They have adopted Programmable Automation and Control technology. Fuzzy Logic was used to identify the type of fault
20. Widodo DA et al (2021) Renewable energy power generation forecasting using deep learning method. In: The 9th engineering international conference, IOP conference series: earth and environmental science, vol 700, pp 012026. https://doi.org/10.1088/1755-1315/700/1/012026. file:///C:/Users/J%20NAMRATHA%20MANOHAR/Downloads/Renewable_energy_power_generation_forecasting_usin.pdf

Chapter 6
Modern Technologies in Grid Integration of Renewable Energy

6.1 Introduction

Driven by both environmental and economic benefits, the Key trends towards discovery and development of modern technologies, include digitization for streamlined operations and energy-efficient integrations to tackle intermittency issues. Advances including AI-enhanced grid management [1] and next-gen battery storage, complement untapped water energy sources like tidal, wave, and ocean currents and offer green energy solutions for tomorrow. The top 10 Green Energy Innovation Trends (2025) [1] identified, by data collection and applying Big Data Analytics are:

1. Advanced Photovoltaics
2. AI and Big Data
3. Distributed Energy Storage Systems
4. Hydropower
5. Wind Energy, Wave Energy
6. Bioenergy
7. Grid Integration
8. Green Hydrogen
9. Advanced Robotics
10. Blockchain

In this chapter we shall discuss about wave energy and storage technologies.

6.2 Wave Energy

The Wave Energy Conversion System (WECS) consists of four stages [2]:

1. Absorption,
2. Transmission,

N. Manohar et al., *Machine Learning Applications in Renewable Energy*, Green Energy and Technology, https://doi.org/10.1007/978-981-97-9939-8_6

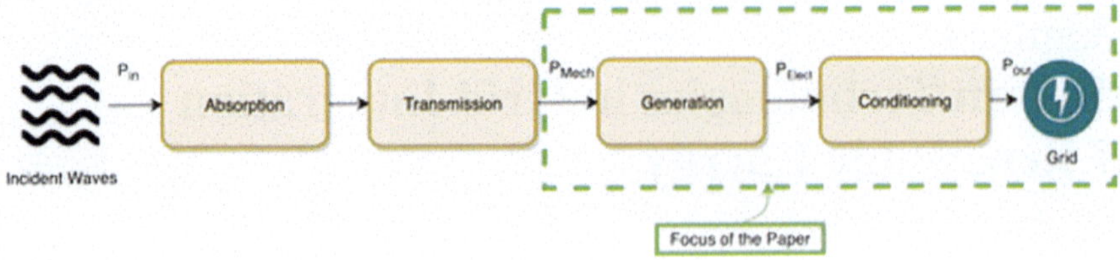

Fig. 6.1 Stages of WECS

3. Generation and
4. Conditioning, as shown in Fig. 6.1.

Each stage of the conversion system has its dynamics and constraints.

Absorption stage is the conversion of wave motion into oscillating motion of the WECS. This absorbed energy is transmitted into hydraulic energy in the transmission stage and converted into electrical energy in the generation stage. Finally, the outgoing power signal is adapted to be delivered into the grid in the conditioning stage. In some particular cases, known as direct conversion, the WEC is directly connected to the electrical generator using a linear generator, in which case, the transmission stage is eliminated resulting in a three-conversion-stage system (absorption-generation-conditioning.

Models that incorporate all these stages from waves to the grid are known as wave-to-wire (W2W) models. Various wave energy devices have distinct characteristics with respect to grid connection, due to their different power output profiles which, in turn depends on inertial characteristics of each device. The inertial characteristics of a generating plant can be defined as that property of the generator mass which overcomes an immediate imbalance between supply and demand. The classification [3] of WECS can be based on working principles, or on a location basis, as shown in Fig. 6.2. For grid integration studies, it is essential to classify WECs on the basis of their inertial characteristics, and a device with a more constant power output will be easier to connect to the grid, since less power conditioning is needed.

The power output of wave energy devices is not ideal, due to the variability of the wave resource, and it cannot be directly connected to the grid in a traditional manner. Grid integration process encompasses various problems ranging from suitable control of power converters involved and power conditioning processes.

Hence, the use of an energy storage [4] component for grid integration of wave energy allows increased utilisation of this intermittent source by mitigating the effects on the power network and also provide ancillary services to future grids.

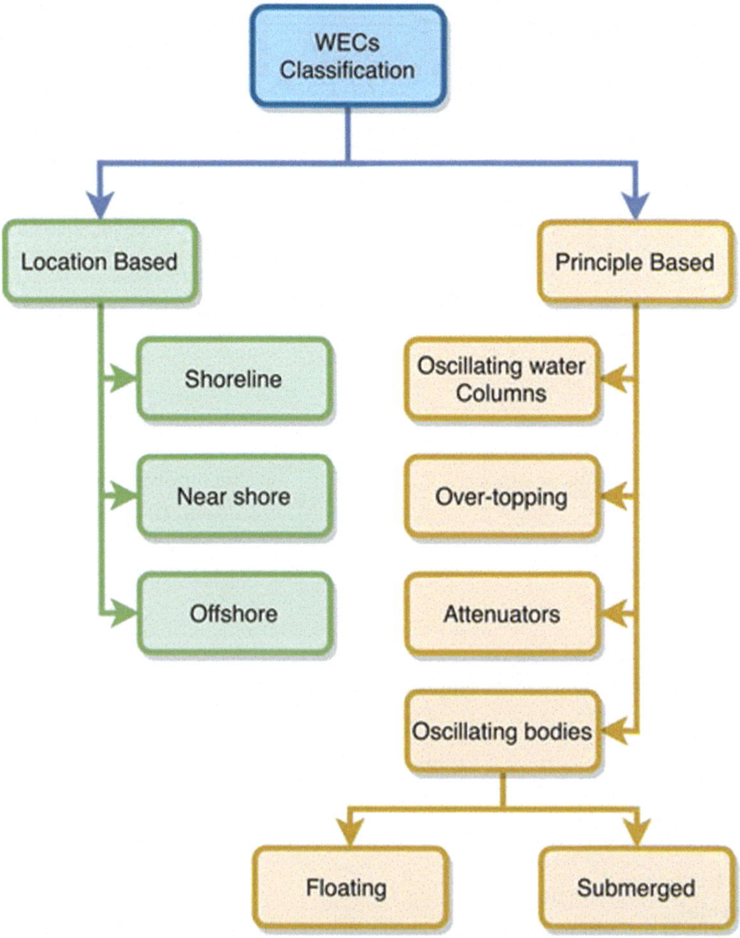

Fig. 6.2 Classification of WECS

6.3 Energy Storage Technologies

Significant advancements have been made in renewable energy storage technologies, which play a crucial role in the development, operation and maintenance of grid integration market. Energy storage systems, such as batteries, pumped hydro storage, and advanced flywheels, enable the capture and store excess energy generated during periods of high production. The stored energy is released during times of low energy generation, which ensures a continuous power supply and enhances grid stability.

There being several options for storage devices, the proper selection of these devices for a particular application is vital. With regard to grid integration, the energy storage system performs:

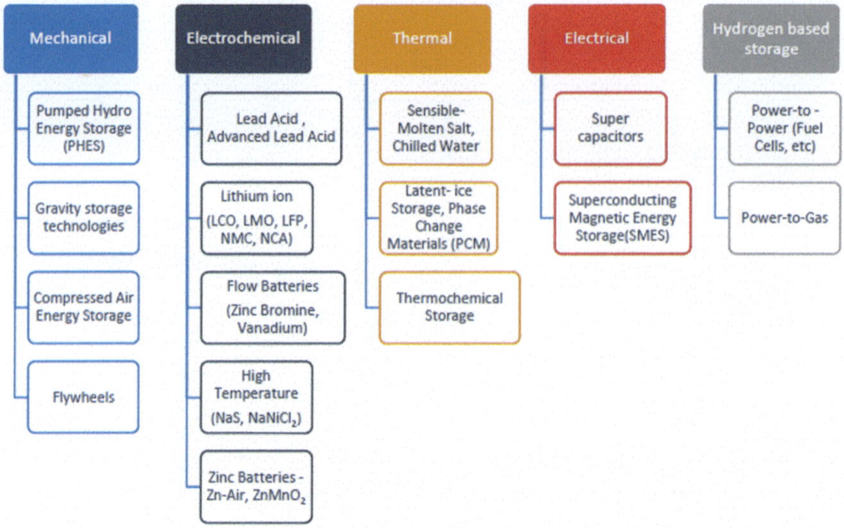

Fig. 6.3 Types of renewable energy storage systems

- Smoothing of output power,
- Short duration fluctuation suppression and helps in frequency and voltage regulation.

Renewable energy storage systems is presented can be broadly divided into five categories, as in Fig. 6.3: The categories are:

1. Mechanical
2. Electrochemical
3. Thermal
4. Electrical
5. Hydrogen based storage

6.3.1 Mechanical Energy Storage Systems

Mechanical energy storage systems generally have higher power ratings due to the size of the storage component. Mechanical storages are:

- Pumped hydro storage
- Gravity storage technologies,
- Compressed Air Energy Storage (CAES) and
- Flywheels.

For effective grid integration, a fast response time and high energy density are required. Flywheels might be best suited for this purpose among mechanical energy storage technologies, taking into consideration the drawback of high self-discharge levels.

6.3.2 *Electrical Storage Technologies*

Electrical storage technologies include

- Capacitors,
- Super-capacitors and
- Superconducting Magnetic Energy Storage (SMES).

Higher cycle efficiency and peak power deliverability of these devices make them suitable for grid integration and power quality improvement applications, despite their poor self-discharge response.

6.3.2.1 Thermal Energy Storage Systems

Thermal energy storage systems are

- Aquifer low-temperature,
- And high-temperature thermal energy storage.

Due to lower efficiency and the early development stage of low-temperature thermal energy storage systems, systems such as cryogenic energy storage systems are not considered to be a good option for grid integration of wave energy despite having high energy density. On the other hand, high-temperature systems such as sensible and latent heat storage are in use, but lower efficiency makes it difficult for use in grid integration technologies.

6.3.2.2 Chemical Energy Storage Systems

Chemical energy storage systems [5] are the most widely used storage systems in various industries, and appliances as mobile phones, electric vehicles etc. They are classified into:

- Conventional batteries and
- Advanced batteries.
- Chemical storage

Conventional batteries are those batteries which are mature technologies such as lead-acid, Ni–Cd and Ni-MH batteries, and are used in a wide range of applications. Advanced batteries include lithium-ion (Li-ion), sodium-sulphur (Na-S) and sodium nickel chloride (NaNiCl) batteries. Li-ion is the most commonly used technology, which can be found in almost every portable electronic device that uses batteries.

Chemical storage include fuel cells [6], which come in various forms and are used in many applications, including grid integration, due to their fast response characteristics [7], the flexibility of power. As a result, they are widely used in grid integration of renewable energy, voltage control and frequency regulation.

6.4 Hybrid Energy Storage Systems (HESS)

HESS is combining the characteristics [8, 9] of two or more energy storage devices. HESS is configured for specific applications such as electric vehicles and grid support. For grid integration of renewable energy generation, the use of HESS is becoming popular. HESS usually combines one high peak power device and one high energy device, for example, battery and super-capacitor HESS, or fuel cell and super-capacitor HESS. Both configurations use batteries and fuel cells as high energy devices, and super-capacitors as the high peak power devices. The quick response of super-capacitors and the high energy density of batteries makes such a combination very useful for grid integration studies. In grid integration, high peak power is required to meet sudden load peaks and high energy density is required to store energy in bulk. Generally, these are combined using the following three topologies:

- **Passive topology**: In this topology, both sources are connected in series or parallel without any power converter to control the flow of power. This is the simplest topology, but there is no control over the power distribution between the sources.
- **Semi-active topology**: One of the devices is connected through a power converter while the other one is connected directly. This topology allows some level of control over power distribution among the sources.
- **Active topology**: A fully active topology uses power converters for both storage components of HESS and through the control of power converters. This topology allows full control of power distribution among the sources.

6.4.1 *Choice of Storage Option for Wave Energy Integration*

For grid integration of wave energy, the use of energy storage systems [10, 11] is primarily for power quality improvement and frequency regulation. Modern renewable energy resources do not provide inertia support to the grid like conventional power plants. This creates an opportunity for energy storage systems to emulate this inertia response for renewable energy generating devices.

Grid integration studies of WEC devices can be categorised based on the storage component used:

- Grid integration of WEC with no storage component
- Grid integration of WEC with one storage component
- Grid integration of WEC with HESS

Most grid integration studies use a DC link capacitor as short term storage, between back-to-back power converters.

6.5 Impacts of Wave Energy on Existing Grids

Power quality is one of the main issues considering the impact of wave energy or any other renewable generation on the grid, due to the intermittency of the resource. Increasing distributed generation, and their contribution to the grid, significantly affects the ability of the power system to deliver good quality power. Because these generating stations do not usually contribute to tasks that conventional generating stations do, e.g. voltage control through reactive power, frequency control through active power control and support during grid faults. Although grid codes are now updated to incorporate these changes for wind power plants, the same will follow for wave energy as its contribution to the grid increases.

The selection of the point of grid connection will depend upon the capacity of the wave device or wave farm. For a single device, it is expected to have a connection with the distribution network rather than a high voltage transmission line connection, and its effect on the grid will not be significant enough to influence voltage and frequency control. However, its impact on the local distribution grid, in the form of small voltage variation such as harmonics induced by power converters, and flicker, may be observed. If the local grid is weak, then the flicker amplitude will be higher and can cause problems for consumers, but can be mitigated by the use of filters and energy buffers. Under grid faults, the contribution from a single WEC device is not significant and cannot contribute to fault currents more than the power converter's rating. However, the use of an external static synchronous compensator (STATCOM) can help in compensation.

In the case of wave farms connected to the grid, the first thing to consider is the connection point which will, most likely, be on the high voltage transmission system, rather than a distribution network. It is expected that they provide voltage and frequency support similar to other generating plants. Voltage source converters can provide voltage support through reactive power control. However, frequency control needs active power control, which is complicated to perform for wave energy device because the resource is highly variable and needs storage as a buffer or a backup. One way to provide this support is to design controllers.

6.6 Combinational Studies

In some situations, wave energy has potential benefits over wind and solar such as better predictability and lower environmental impacts. These benefits, and advances in offshore wind technology, have led to positive synergies between wind and wave. The data from these studies suggests that it is advantageous to use these technologies in combination [12]. In this regard, several studies have been conducted in the past decade, which describe the temporal complementarity between the two resources. Solar and wave energy have been used in a study for island communities in Malaysia and the results are encouraging. The use of a mix of renewable energy sources reduces the reserve requirement, compared to a single renewable energy source. This allows for more renewable penetration into the existing mix and emphasizes the need for combined studies.

Wave energy can be used with either wind or solar energy depending upon the particular site characteristics. The main obstacle is the relatively low technology readiness level of wave power compared to wind and solar. Lack of experience in the field of wave energy and higher insurance costs are also contributing factors to the slow development of wave energy devices. In addition, it is necessary to develop a (costly) full-scale model for combined studies and detailed analysis of the cost and performance of the system.

6.6.1 Wave and Solar

The solar energy resource [13] is quotidian and has seasonal variability with the additional variability due to cloud cover. In contrast, the wave resource is generally smoother with the exception of storm events, particularly during the winter season. For this reason, wave energy can be used in conjunction with solar, especially on remote islands with an abundance of sunshine in the day time. In [14], the authors explain the importance of adding wave energy to solar energy in Karnataka, India, in terms of carbon emissions. It is concluded that the wave-solar hybrid combination is most effective in Monsoon seasons when solar resource is minimal. The additional use of storage further enhances the combination.

6.6.2 Wave and Wind

The idea of combining wave and wind energy [15] technology is prudent, considering the rapid development of offshore wind energy technology. Combining offshore wind and wave can produce the following key benefits:

- **Reduced power variability**: Renewable sources produce variable power due to various environmental conditions. Using complementary technologies like offshore wind and wave reduces the variability of power produced.
- **Better predictability**: The predictability of wave power is, in general, better than the wind energy. For example, one study shows that the predictability of wave energy is 23% better than wind, leading to better power output predictability for a wave energy converter (35% better than wind).
- **Increased the amount of Power Output generated**

- **Shared costs**: These costs may include structural foundation costs, operations and maintenance (O&M) costs, balancing costs, shared grid infrastructure costs, underwater transmission costs and reduced environmental impacts.
- **Efficient** utilization of offshore site areas.
- **Streamlining of maintenance and grid connections** [16].

The prospect of using the same marine space and lowering overall environmental impacts on marine life is an additional benefit. There are various studies on combined wave and wind resource harnessing using multiple stochastic and deterministic approaches. They describe the importance of using combined technologies, including increased power capacity and reduced output power variability. Combined wave and wind energy can be exploited in two ways:

(1) In combined farms/arrays, but on separate platforms, and
(2) By combining both technologies on the same platform, especially for offshore wind and wave.

Combined wave–wind systems can be classified based on:

i. Technology,
ii. Water depth as shallow, transition or deep water,
iii. Location relative to the shoreline as shoreline, nearshore, offshore.

Pérez-Collazo et.al in their work [17] have proposed a new classification based on the degree of connectivity between offshore wind turbines and WECs as:

(A) **Co-located**: Co-located independent arrays are those which, while constituting distinct offshore wind and wave farms and occupying different marine areas, are close enough to share the same electric grid connection alongside other services or installations.
(B) **Hybrid**: Hybrid systems are part of the wider family of multipurpose-platforms, i.e., offshore structures on which different marine uses are combined, such as: ocean energies, offshore wind, aquaculture, transport and marine leisure. According to their substructure, hybrid systems can be classified into bottom-fixed and floating, appropriate for shallow or transitional and deep water, respectively
(C) **Island systems**: The third and last family of combined wave and wind energy technologies is island systems. Island systems are offshore multipurpose platforms. The differences with hybrid systems lie in that islands tend to be much larger and, perhaps more importantly, unify the combined exploitation of more than two marine resources on the same platform. Island systems can be divided into artificial or floating islands.

6.7 Value of Wave Energy on the Grid: An Economic Aspect

6.7.1 Cost of Wave Energy Plant

Evaluating the economic feasibility of any power generating plant is one of the impotant planning and operation parameter. The cost of a wave energy plant can be evaluated using the levelized cost of electricity (LCoE). LCoE is a metric that combines all the costs associated with a wave energy plant, including:

- Capital expenditure
- Operational expenditure
- Decommissioning costs
- Annual energy production
- Discount factor
- Project lifespan

LCoE is a useful metric because it consolidates all the direct costs of a wave energy plant into a single value. This makes it a common way to compare different methods of electricity generation.

Other costs to consider when evaluating a wave energy plant include:

- **Pre-operating costs which** includes the costs of environmental impact assessments, preliminary studies, consenting procedures, and more. The value of pre-operating costs can vary depending on the location, installation type, and other factors.
- **Cost variability**: Wave energy technologies can have significant cost variability due to the many different prototypes, anchoring and fixation systems, and energy transport systems.
- **Economic drivers**: The primary economic drivers for wave energy plants are geographical variations, durability, and power demand.

References

1. https://www.startus-insights.com/innovators-guide/top-10-renewable-energy-trends-2022/
2. Penalba M et al (2016) A review of wave-to-wire models for wave energy converters. https://www.mdpi.com/1996-1073/9/7/506
3. Said HA, Ringwood JV (2021) Grid integration aspects of wave energy—overview and perspectives. https://doi.org/10.1049/rpg2.12179
4. Kumar M, Shanmugam K, Pradeep K, Fillippone M et al (2022) Grid integration and application of battery energy storage systems. https://ieeexplore.ieee.org/document/9865726
5. https://www.blackridgeresearch.com/blog/all-you-need-to-know-about-an-energy-storage-system-ess-components-technologies-benefits-risks-applications-bess-battery-energy-storage-system
6. https://www.eia.gov/energyexplained/hydrogen/use-of-hydrogen.php

7. https://www.plugpower.com/fuel-cell-power/fuel-cell-benefits/
8. https://www.frontiersin.org/journals/energyresearch/articles/10.3389/fenrg.2022.990653/full
9. Zhuge K, Kazerani M (2014) Development of a hybrid energy storage system (HESS) for electric and hybrid electric vehicles. IEEE. https://ieeexplore.ieee.org/document/6861868
10. Said HA, Ringwood JV (2021) Grid integration aspects of wave energy—overview and perspectives. https://doi.org/10.1049/rpg2.12179
11. Pelosi D et al (2024) A hybrid energy storage system integrated with a wave energy converter: data-driven stochastic power management for output power smoothing. https://www.mdpi.com/1996-1073/17/5/1167
12. https://link.springer.com/article/10.1007/s11802-021-4708-1
13. https://www.sciencedirect.com/science/article/abs/pii/S0196890420308384
14. Goswami P, Deshmukh SP (2018) Assessment of reduction in carbon dioxide emission with wave solar hybrid generation along coastal Karnataka. Nat Environ Pollut Technol Int Q Sci J 17(3):863–868. ISSN: 0972-6268. https://www.neptjournal.com/upload-images/NL-65-26-(24)B-3505.pdf
15. Chozas JF Kofoed JP Sorensen HC (2013) Predictability and variability of wave and wind: wave and wind frecasting and diversified energy systems in Danish North Sea. DCE Technical Reports No.156. Department of Civil Engineering, Aalborg University
16. McTiernan KL, Sharman KT (2020) Review of hybrid offshore wind and wave energy systems. J Phys: Conf Ser. https://doi.org/10.1088/1742-6596/1452/1/012016/pdf
17. Pérez-Collazo C et al (2015) A review of combined wave and offshore wind energy. Renew Sustain Energy Rev 42:141–153. https://doi.org/10.1016/j.rser.2014.09.032

Chapter 7
Big Data and Machine Learning for Hybrid Power System—Power Quality

7.1 Introduction

Increased digitization, rising complexity and expansion of power systems, has resulted in significant increase in the quantum of data ranging from KiloBytes (KB) to PetaBytes (PB). The branch of statistics has helped us to process the big data to analyse draw inferences and take further action for effective increase in efficiency of power systems. The advent of computer softwares as MS-Excel, COBOL, DBase, SQL has enabled process more big data fast. However, with the increase in automation and deployment of several types and quantity of sensors, power systems becoming closed-loop and need for processing the data and taking corrective action very fast as adjusting power system parameters, the modern techniques as Big Data (BD) Analysis, Machine Learning Algorithms (MLA) support energy systems to enhance the technical, operational, economic, and environmental benefits. The *volume* of data being generated in the power systems [1] has grown tremendously over the past few years and also are generated by *variety* of sources, locations, types and applications. Figure 7.1 shows the various sources of Big Data in power systems. Additionally, many forms of grid data are generated at high *velocities*.

The data recorded by many power system sensors, the majority of data is either *not logged*, or they are *overwritten* very quickly. For example:

- In most protection relays and related sensors, the data collected is discarded shortly after internal use, due to which studies as trend analysis is not possible.
- In power quality sensors data is stored in the form of the voltage or current waveforms only a few cycles before and after an *event* is detected.
- Most measured data in power systems are intended to be used close to the point of their generation, and were never transferred to an central data center.
- Though many of the recently deployed or emerging power data measurement systems lie in the description of BD, the they are currently managed does not exactly match the *soul* and *purpose* of BD. Once such hidden data is collected, managed, and analyzed, they will constitute the real BD in power systems.

N. Manohar et al., *Machine Learning Applications in Renewable Energy*, Green Energy and Technology, https://doi.org/10.1007/978-981-97-9939-8_7

Fig. 7.1 Various sources of big data for power systems

Machine Learning is an Artificial Intelligence Technique in the field of computer science. It is the study of automated data processing or decision-making algorithms that improve themselves automatically based on experience or past experience. The main objective of machine learning model is to develop computer programs that can access data and use it for learning purposes, to constantly improve system performance. With the advancements in Big Data, Machine Learning[2] has become a key player in solving problems in various areas such as:

- Image recognition
- Speech Recognition
- Healthcare
- Finance and Banking industry
- Computational Biology
- Energy production
- Automation
- Self-driven vehicle
- Natural Language Processing (NLP)
- Personal virtual assistance
- Marketing and Trading
- The education sector, etc.

All the above areas mentioned are an integral part of power energy systems (PES). The integration of machine learning with power energy systems transforms PES into smart energy systems [2]. Machine learning-powered devices when incorporated into existing energy grids, makes the systems capable of real-time adjustments for fluctuating energy supply and demand thus making the hybrid power system very stable.

BD and MLA continuously analyzing and optimizing energy distribution, thus making energy grids more resilient, adaptable, and capable of effectively handling the challenges posed by the growing diversity of energy sources and demands.

It may be noted that the renewable energy system in the context of this book is a hybrid power system with many energy sources and is also a smart grid. Hence, the terms renewable energy system, hybrid power system and smart grid will be used based on the application concept.

7.1.1 Applying Machine Learning and Big Data for Various Renewable Energy Sources (RES) Power Systems

In this section we shall study the various issues and problems that can be tackled efficiently by applying machine learning and data analysis in various RES power systems.

Wind Energy: Wind turbines use a multitude of sensors to measure ambient and dynamic performance variables such as humidity and temperature, and turbine status such as vibration and speed etc. These parameters are measured by supervisory control and data acquisition (SCADA) system to monitor the system's condition. Machine learning algorithms [3], especially ANN, are widely employed to process the data, for aspects such as:

i. Fault detection of hardware components as gear box bearings, generator rotor and stator, turbine blades, inverters etc.
ii. Forecasting of wind speed and power [4].
iii. Optimization of wind turbine design.
iv. Control of maximum power tracking, pitch angle, speed, mechanical and electric power, torque etc.
v. Classification and analysis of faults.

Solar Energy: Continuously monitoring and analysis of various parameters of solar radiation environment, solar application system—such as temperature, voltage, and current, machine learning algorithms [5] can identify under performing panels or potential faults. This enables improved performance as maintenance efforts can be directed precisely where they are needed, preventing energy loss due to sub-optimal panels. Additionally, machine learning techniques contribute to:

- Solar output forecasting by predicting cloud cover patterns. These predictions provide critical insights into expected fluctuations in energy production, enabling grid operators to balance supply and demand more effectively.
- Solar system efficiency by monitoring Solar Irradiation, weather conditions, load demand.
- Optimize Plant layout

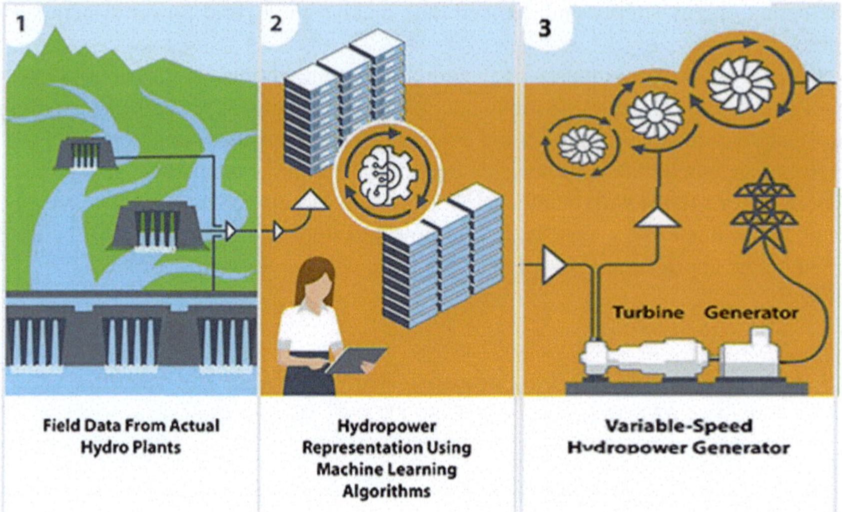

Fig. 7.2 Machine learning applied for hydroelectric generation power system

- Minimize Shading for increased output
- Better Operations and maintenance.

Hydroelectric Energy: Fig. 7.2 **depicts the application of machine learning algorithms (MLA) in hydroelectric energy power system** [6]. **The stages of hydroelectric power generation or any renewable energy system are:**

1. **Field data is collected from the operational hydroelectric power plant.**
2. **The data is then organized in the standard format for easy application of machine learning algorithms.**
3. **The decision analysis of the machine learning and data analytics is sent to the action taking centers of the power plant for necessary adjustments of the operational parameters, to optimize the output.**
4. **The hydroelectric plant operates with the adjusted inputs.**

Machine learning enables predicting water levels in reservoir accurately by integrating the analyses outcomes of historical data, weather forecasts, and hydrological patterns The resultant information supports taking decisions for optimizing energy production by adjusting water release strategies to match energy demand. Machine learning not only supports dynamic power production optimisation but also aids in the preventive maintenance of hydroelectric infrastructure. By analyzing sensor data from turbines, generators, and other components, potential issues can be identified early, preventing costly breakdowns and ensuring continuous, efficient operation.

vi. **Energy Grid Management**: Machine learning offers substantial benefits to energy grid management [7]. Steps for application of MA and BD for Energy Grid Management are:

1. Various sensors record data.
2. The recorded data is transmitted to the Central Computing System (CCS).
3. At the CCS the data is initially refine to eliminate noise.The refined data is analysed for the outcome.
4. Data obtained in step 3 along with historical consumption data, weather conditions, and even socio-economic factors are processed through Machine learning models analyze to

- predict energy demand.
- Help grid operators to allocate resources more effectively and avoid energy wastage.
- Improve grid stability.
- Identify unusual patterns or potential faults, enabling swift intervention and prevention of disruptions.
- Make accurate predictions, and adapt to changing conditions for RES to operate more efficiently, reduce downtime, and contribute to a greener, more sustainable future.

7.1.2 Benefits of Integrating ML with RES

Integrating ML with RES offers several advantages, significantly enhancing their efficiency and sustainability [8]. These benefits applied to various aspects as technical, financial and environmental are:

- **Increased Energy Production Efficiency**.
- **Harness** the maximum potential of RES.
- Early detection of potential equipment malfunctions, by predictive maintenance algorithms.
- Minimize downtime.
- Optimized maintenance schedules.
- Reduce operational costs.
- Ensures a consistent energy supply.
- Reduce various technical losses.
- **Enhanced Resource Management**.
- Reduce the environmental impact of renewable energy generation.
- Adjust operational parameters to meet energy demands
- **Transition to Smart Energy Systems**.
- Continuously analyzing and optimize energy distribution, making energy grids become more resilient, adaptable, and capable of effectively handling the challenges posed by the growing diversity of energy source.

Thus, the benefits of integrating ML into RES extend beyond mere technological enhancements.

7.1.3 Challenges and Limitations of Integrating ML with RES [9]

- Selection and deployment of sensors.
- Embedding MLA and BDA in existing infrastructure computing systems.
- Selection of appropriate MLA. The selection has three phases-model learning,model training and verification and third phase is model deployment.
- Data Privacy and Security Concerns.
- Interoperability.
- Scalability and Implementation.
- Handling of sensitive energy consumption data, as there is an inherent risk of compromising user privacy. Balance between utilizing comprehensive data and protecting individuals' privacy is important.
- Increasing interconnection of renewable energy systems calls to address cyber threats,causing disruption of energy production, distribution, which may cause damage to loads and customer resources.
- Gaining trust of Regulatory bodies, investors, and even the general public who may be hesitant to embrace technologies they cannot fully comprehend with.

7.1.4 Concepts of Big Data

The term big data was first used by John Mashey to refer to handling and analyzing massive datasets. Doug Laney defined big data by the "3V's": volume, velocity [10], and variety. Present day Big data is defined by the "4V's" by adding veracity to the "3V's".

(a) *Volume*: Volume is the amount, size, and scale of the data. When in any system the amount of data reaches a level which cannot be managed by simple analytical tools, then it can be easily handled by Big Data Analysis. The size defines two aspects of the database-one vertically by the number of samples or records in a data set and other horizontally by the number of features.
(b) *Velocity*: Velocity refers to the speed at which the data is generated in a system and how fast the data should be processed for effective control of the system.
(c) *Variety*: Variety is the heterogeneity of the data. Big data collected in a system often comes from different stages and sources, which care diverse in type, format, semantics, and volume.

The so called "three Vs" of BD, the measuring units and the source of generation of each characteristics is depicted in Fig. 7.3.

The fourth V of big data is:

(d) *Veracity*: Veracity refers to the quality of the collected data. It considers biases, noise, and abnormality in data. The accuracy of any analytic process applied to the data depends greatly on the veracity of the source data.

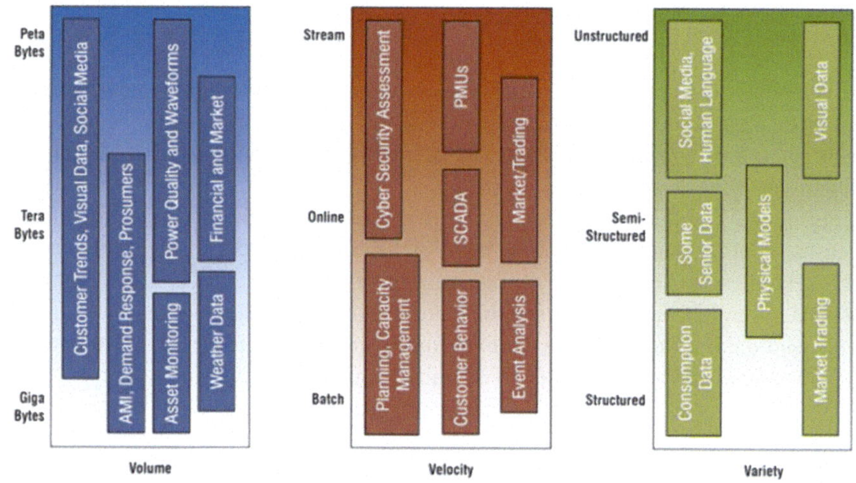

Fig. 7.3 The 3 V attributes of BD and some examples in the context of power systems

7.1.5 Power Quality (PQ)—Data Recording

The term "PQ of big data" refers to the quality of the large amount of data resulting from continuous PQ monitoring [11] in a power system. Power grid stakeholders continuously monitor Power System (PS) data to obtain information on PS performance. Long-term PQ measurements result in a large amount of data in several Giga Bytes (GB).

The basic parameters measured in a power system, to analyse [12] the performance state of the power system and PQ monitoring are voltage and current samples in the time domain. The PQ BigData processes the raw data of the voltage and current samples, to detect and extract events such as-

- Power factor
- Voltage variations—as sag and swell
- Transients
- Waveform distortion
- Power flow at various sections and stages of the power system.

Raw data is the measured value or data received from sensors. Raw data is not organized and may contain some noises such as distorted signals. The raw data is processed through modulation techniques to obtain refined data. Refined data is then subjected to BD and MLA to obtain decision strategy.

7.1.5.1 Volume of PQ Data

V, volume, is the amount of data that is collected and analyzed. Smart meters installed in homes are just one example of the data sources and the scale of data collection that such devices generating, about 120 GB of data. Big data applications, can handle data in the order of terabytes and petabytes, from various applications as mentioned in Fig. 7.1. PQ monitoring generates a large amount of data. For instance, waveform measurements contain 256 samples per cycle, considering a sampling frequency of 12.8 kHz in 50 Hz. For each hour of measurement, this results in 10,80,000 samples (3 voltages and 3 currents). This volume complexity also holds for the raw data and processed data. To consider an example harmonic values are obtained by Fourier transformation and aggregated every 10 min to simplify the data analysis.

7.1.5.2 Velocity of PQ Data

The second V, velocity, is the speed of data generation and transmission of that data. The velocity of PQ data is related to the set sampling frequency in the PQ monitor. The frequency of data collection, that is the number of samples collected per unit time, directly affects the volume of data that is collected and that needs to be stored. The higher the frequency more is the data. Metering units and sensors have varying sampling rates, which range from readings at near-constant frequency to event-triggered readings, Considering the previous example of 256 samples per cycle, which means a new sample in every voltage and current channel is obtained every 78.125 μs. Utility devices such as PMUs have sampling rates as high as 240 samples per second, while smart meters and satellite weather data can have as high as 15 min between measurements. The velocity complexity is also related to the PQ data analysis streams: whether done offline and online analysis. Offline analyses are mainly suitable for system performance evaluation, problem characterization, and systems diagnosis. For online analysis, results are convenient when actions must be taken immediately.

Various data optimization, compression, and searching methods for more advanced data analyses are crucial for enabling higher collection rates. Once thedata is collected, parsing through it quickly to classify and identify meaningful data amongthe collected data falls into the realm of advanced data mining processes.

7.1.5.3 Variety of PQ Data

The third V, variety, is to the range of data that are collected and utilized. Though the raw PQ data comes from only two signals -voltage and current, they are obtained from several measurement sources as intelligent relays, smart meters, digital fault recorders, phasor measurement units (PMU), etc. PQ monitors are installed at distinct locations of a power system to obtain enough data to understand and characterize the PQ phenomena. While greater variety of data facilitates greater meaning, processing this semi –structured and un-structured data is challenging for systems and methods, which is possible through big data analysis. All information collected under various formats are processed by sorting and converting to usable structures for processing and decision making.

Depending on the monitor settings, harmonic values and THD can be obtained each 10 min, voltage dips and transient waveforms at a frequency of 256 samples per cycle, and rms voltage over 150 cycles of the power-system frequency. Although there are standardized methods to extract the data, as IEC 61000-4-30, PQ monitors can also be configured with different settings, which results in data with different sampling, formats, and sizes. To reduce this complexity, the IEEE 1159.3 PQDIF Task Force has developed a standard format called Power Quality Data Interchange Format (PQDIF).

7.1.5.4 Veracity

Fourth V, Veracity refers to the accuracy and reliability of the data. Large collection of data have higher scope for inaccuracies due to measuring errors, equipment tolerances, or transmission noise. For example, smart meters, have as much as a 2.5% measurement error. While this alone is not a particularly high error, a system reliant on the readings of millions of smart meters, each with a 2.5% error, may not be able to optimally process and then respond to network states.

According to the standard IEC 61000-4-30, PQ measurement devices must comply with specific accuracy requirements (class A requirements). In particular, for harmonics, accuracy requirements are defined in IEC 61000-4-7. This accuracy is influenced by many factors, which include external sensors (e.g. instrument transformers, Rogowski coils), A/D converters, or measurement algorithms (e.g. aliasing, leakage). The requirements for PQ measurements point out the veracity complexity of PQ data.

7.1.6 Big Data Analytics

Big data analytics is the process of extracting information and detecting patterns from datasets with the 4V's complexity. PQ monitoring is the process of gathering, analysing, and interpreting voltage and current measurements into useful information. The continuous PQ monitoring allows the network operators to obtain information about the performance of utilities and customer facilities.

Analysis of PQ variations over time is essential to identify, for instance, excessive waveform distortion on power systems as a function of load and system characteristics (e.g. resonance conditions).

7.1.7 Deep Learning (DL) Workflow for Power Quality

The steps in applying DL to a PQ problem in a power system as seen in Fig. 7.4 are:

i. Data Collection
ii. Data Preprocessing
iii. Model selection
iv. Model training
v. Model validation
vi. Model testing
vii. Model improvement.

Step-1 Data Collection: The first step in any DL-based process is to collect data. Regarding the PQ event classification, data can be collected from various sources, such as phasor measurement units (PMUs), smart meters, and power quality monitors. The nature of data in the PQ analysis is generally non stationary and in a discrete time-series form of voltage or current waveforms. It comprises information on the PQ event's type and duration.

Step-2 Data Preprocessing: the collected data may contain noise, missing values, or other inconsistencies. Therefore, it is essential to preprocess the data before training the model. This step includes many preprocessing stages, such as data cleaning, normalization, re-sampling, and feature extraction.

Step-3 Model Selection: the next step is to select a suitable model, which depends on the nature of the task to perform with the associated datasets. Several models, such as recurrent neural networks (RNNs), convolutional neural networks (CNNs), and transformer-based models, can be used.

Step-4 Model Training: once the model is selected, the model is trained using the preprocessed PQ datasets. The training process involves feeding the model with the training input datasets and optimally adjusting the model's parameters to minimize the loss function. The loss function acts as an index during the training process and measures how well the model performs, and the training process aims to minimize it.

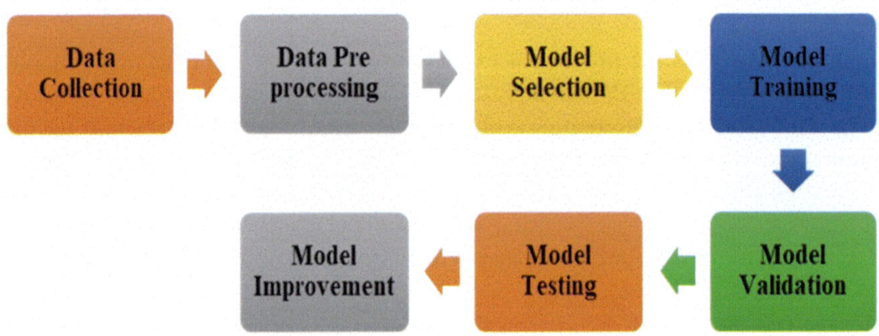

Fig. 7.4 Steps to apply deep learning

Step-5 Model Validation: the model's performance is evaluated on a separate validation dataset after the training. This step helps to avoid overfitting, where the model performs fine on the training data and fails to perform better on the new data. The validation dataset should represent the test data and not be used during model training.

Step-6 Model Testing: once the model is trained and validated, it is ready for deployment. The model can classify PQ events in real-time or batch mode. The model outputs either a probability distribution over the different PQ event classes, or a binary decision that may represent the presence or absence of a specific PQ event.

Step-7 Model Improvement: finally, the model's performance is improved by fine-tuning the model parameters, adding more training data, or using ensemble models. The model's performance should be monitored periodically to ensure its relevance to the dynamic state of the system.

7.1.8 Application D Machine Learning and Big Data

7.1.8.1 Case Studies on Implementation of BD in RES

Case Study-I

Curtailing Energy Loss Due to Theft by BD [13]

The common methods of electricity theft are:

- **Bypassing-illegal tapping of electricity from the feeder, as in** Fig. 7.5.

Fig. 7.5 Illegal tapping of electricity from feeder

- **Meter tampering—by grounding the neutral wire as it does not measure readings and**
- **Physical methods to evade payment of bills.**
- **Direct wire-connection to a main power route passing a shop or a house so that electricity can flow to the consumer without crossing the electric meter installed by a government agency which is responsible for providing electrical services to customer.**

There are different types of theft done all over the world. Huge amount of power theft is done by tapping from line or bypassing the meter, According to a study 80% of the total theft detected all over the world is from residential buildings and 20% from commercial and industrial premises.

BDA techniques can be utilized to curtail thefts by comparing standard usage patterns of customers. Trained models can detect the difference between normal usage pattern and unusual usage pattern. The electricity consumption changes frequently and a large amount of installed IoT devices monitor the multi-source real-time data, such as climatic factors (wind, solar, temperature), transmission and the consumers' electricity usage record. For example, during the uncertain times of COVID-19, when people could be spending more time indoors, the quantity of historical data is big and difficult to analyse. Machine learning (ML), deep learning (DL) and time-series models are the main approaches for electricity theft detection (ETD) in smart grid. Based on smart meter data, normal and abnormal power consumption patterns and footprints can be identified with irregular, longer and higher electricity usage patterns than regular and normal consumption.

References

1. Akhavan-Hejazi H, Mohsenian-Rad H (2018) Power systems big data analytics: an assessment of paradigm shift barriers and prospects. Energy Rep 4:91–100. https://www.sciencedirect.com/science/article/pii/S2352484717300616
2. Josephine NMM, Nathan D (2016) Curtailing energy theft by remote monitoring case study: University of Nigeria In: Proceedings of the world congress on engineering and computer science 2016 Vol I WCECS 2016, October 19–21, 2016, San Francisco, USA
3. Le TT, Paramasivam P et al (2024) Unlocking renewable energy potential: harnessing machine learning and intelligent algorithms. Int J Renew Energy Develop https://ijred.cbiore.id/index.php/ijred/article/view/60387
4. Koti FE, Khajehpoor A (2021) Application of machine learning for wind energy from design to energy-water nexus: a survey. Energy Nexus 2, December 2021, https://www.sciencedirect.com/science/article/pii/S2772427121000115
5. https://arxiv.org/pdf/2303.07875
6. Bernardes J, Santos J et al (2022) Hydropower operation optimization using machine learning: a systematic review. https://www.mdpi.com/2673-2688/3/1/6
7. Sandeep Kumar Reddy V et al (2023) Smart grid management system based on machine learning algorithms for efficient energy distribution. E3S Web Conf 387:02005. https://doi.org/10.1051/e3sconf/202338702005,ICSERET-2023

8. The role of machine learning in renewable energy: enhancing efficiency and sustainability. https://iabac.org/blog/the-role-of-machine-learning-in-renewable-energy-enhancing-efficiency-and-sustainability,2024
9. The impact of artificial intelligence and machine learning on renewable energy. https://www.automate.org/ai/news/the-impact-of-artificial-intelligence-and-machine-learning-on-renewable-energy,2024
10. https://www.digitalocean.com/community/tutorials/an-introduction-to-big-data-concepts-and-terminology
11. Gasch E, Meyer J, Schegner P, Schmidt K (2015) Efficient power quality analysis of big data. In: 23rd international conference on electricity distribution Lyon, 15–18 June 2015
12. Samanta IS (2023) A comprehensive review of deep-learning applications to power quality analysis. Energies 16(11):4406. https://doi.org/10.3390/en16114406
13. https://www.bidgely.com/bihar-partners-with-bidgely-energy-theft-press-release/

Chapter 8
Energies in Quantitative Approach

8.1 Introduction

Engineering applications are all quantitative based. Calculations need to be done for designing, development, implementing, maintenance and further enhancement. The design calculations can be done only if we know the requirements. In the context of renewable energy, the requirements are based on the environment conditions as for solar energy irradiance and wind energy wind speed is required. Hence, each renewable energy system has respective various environmental parameters. Knowing the environmental parameters and the energy output we require for a particular application, we can design the appropriate Renewable Energy System (RES).In this chapter we will study the various computational elements of various types of RES, to enable us determine the design pentameters, outputs and efficiency of the RES.

All computations in the present day are being programmed. Hence, in this chapter to encourage students to apply their computer programming skills for solving real time practical application requirement solutions, programs have been written in 'C' language for some solved problems.

8.2 Solar Energy

Solved Problems [1]:

1. **Determine the battery capacity for a load of 1100 W and backup time of 5 h?**

$$\text{Sol}: \ \text{Load} = 1100\,\text{W}$$

$$\text{Back} - \text{Up Time} = 5\,\text{Hrs}$$

N. Manohar et al., *Machine Learning Applications in Renewable Energy*, Green Energy and Technology, https://doi.org/10.1007/978-981-97-9939-8_8

$$\text{Battery Capacity} = \text{Load} \times \text{Back} - \text{Up Time}$$

$$100\,\text{W} \times 5 = 500\,\text{W}$$

2. **Calculate the daily solar energy received by a standard hot water collector of dimensions 1 m by 2 m, in a location which receives 24 MJ/m² day.**

$$\text{Sol : Total area of the panel} = 1 \times 2 = 2\,\text{m}^2$$

Daily solar energy received by a standard hot water collector = 2 × 24 = 48 MJ.

3. **Determine the total area (size) of the solar system with 12 solar panels rated at 100 Watts each and having a conversion efficiency of 18%?**

Sol: The total power output of the solar system can be calculated as:

$$\text{Total Power Output} = \text{Total Area} \times \text{Solar Irradiance} \times \text{Conversion Efficiency}$$

$$\text{Given : No. of solar panels} = 12$$

$$\text{Output per panel} = 100\,\text{Watts}$$

$$\text{Total Output Power} = 12\,\text{panels} \times 100\,\text{Watts} = 1200\,\text{Watts}$$

The Solar Irradiance for a surface perpendicular to the sun's rays at sea level on a clear day is about 1000 Watt/m^2.

$$\text{Conversion Efficiency} = 18\%.$$

Therefore, Total Area or size of the system = **Total Power Output/(Solar Irradiance x Conversion Efficiency)**

$$= [1200/\,(1000 \times 0.18)] = 6.66\,\text{m}^2$$

4. **Determine the solar declination angle on February 15.**

Sol: February 15, day of the year n = 46
The solar declination, in degrees $\delta = 23.45\text{Sin}[360(284 + n)/365]$

$$= 23.45\text{Sin}[360(284 + 46)/365]$$
$$= -13.11^\circ$$

C-Program

```
/* Program to compute declination angle given day of the year */
/*Example of day of the year is February 15, then the input should be 46(31
+ 15)*/
#include<stdio.h>
#include<math.h>
main()
{
int n; /*n = day of year*/
float dangle;
printf("\nEnter day of the year");
scanf("%d",&n);
dangle = 23.45 * sin(0.0175 * 360 * (284+n)/365);
/*factor 0.0175 to convert degrees to radians*/
printf("\nDeclination angle on %d day of the year = %f",n,dangle);
return 0;
}
```

5. **Calculate the solar hour angle at 09.00 and 13.00 solar time.**

$$\textbf{Sol}: \text{Hour angle}(\omega) = 15(\text{t} - 12)$$

At 09.00 Hour angle $(\omega) = 15(9 - 12) = -45^{\circ}$.
At 13.00 Hour angle $(\omega) = 15(13 - 12) = 15^{\circ}$.

C-Program

```
/* Program to compute hour angle */
#include<stdio.h>
main()
{
float t,hangle;
printf("\nEnter time of the day(0.00 – 12.59)");
scanf("%f",&t);
hangle = 15 * (t – 12.00);
printf("\nHour Angle at time %f = %f",t,hangle);
}
```

6. Determine the efficiency of a solar panel **with a maximum power output of 200W and a surface area of 1.6 m^2**.

Sol: The efficiency of solar panels is given by:

$$\text{Efficiency}(\%) = [(\text{P}\max \div \text{Area}) \div 1000] \times 100\%.$$

$$\text{Efficiency} = [(200 \div 1.6) \div 1000] \times 100\% = 12.5\%.$$

In this case, the solar panel efficiency is 12.5%. That said, it can convert 12.5% of the sunlight that hits the panels to electricity.

7. A single solar cell (10 cm × 10 cm) produces a voltage of 0.5 V and a current up to 2.5 a. If the solar insolation is 800 W/m^2, determine the efficiency of the solar cell?

$$\text{Sol : Given Values.}$$

$$\text{S.I} = 800\,\text{w/m}^2 =$$

$$\text{Voltage(V)} = 0.5\,\text{V}$$

$$\text{Current(I)} = 2.5\,\text{A}$$

$$\eta = ?$$

$$\text{Area} = 10\,\text{cm} \times 10\,\text{cm} = 10^{-2}\text{m}^2$$

$$\eta = \left(\frac{Pout}{Pin}\right) \times 100 \quad (1)$$

$$\text{Solar Insulation (S. I)} = \frac{Pin}{Area}$$

$$Pin = 800 \times 10^{-2} = 8\,\text{W} \quad (2)$$

$$\text{Pout} = \text{Voltage(V)} \times \text{Current(I)}$$

$$0.5 \times 2.5 = 1.25\,\text{W} \quad (3)$$

Using (1), (2) and (3) we can deduce.

$$\eta = \left(\frac{1.25}{8}\right) \times 100$$

$$0.1562 = 15.62\%$$

The efficiency of the solar cell is 15.62%.

8. **Calculate the angle of declination for 7th May of a leap year.**

Ans. 7th May in leap year is 128th day of the year.
Therefore n = 128

$$\delta = 23.45 \sin\left[\frac{360}{365}(284 + n)\right]$$
$$= 23.45 \sin\left[\frac{360}{365}(284 + 128)\right]$$
$$= 16.96^{\circ}$$

Therefore, the angle of declination for 7th May of a leap year is δ = 16.96°

9. Designing a solar car with a total roof area for solar cells of 6.4 m^2. Calculate the electrical power available, assuming total cell efficiency of 17% and a constant light intensity of 980 W/m^2.

Solution:

$$\text{Power input} = (\text{light intensity}) * (\text{roof area})$$
$$= (980\,\text{W/m}^2) * \left(6.4\,\text{m}^2\right)$$
$$= 6.3 \times 10^3\text{W}$$
$$= 6.3\,\text{kW}$$

Since $Efficiency = (Power\ output)/(Power\ input)$,

$$\text{Power output} = \text{Efficiency} * (\text{Power input})$$
$$= 0.17 * (6.3\,\text{kW})$$
$$= 1.06\,\text{KW}$$

Therefor, the electrical power output available = 1.06 KW.

10. A solar hot water system receives solar energy at the rate of 10.5 MJ/m^2 per day. If the collector area is 4.8 m^2, collector efficiency is 0.7 and the water volume is 325 dm^3,
(a) Calculate the total water energy gain per day;
(b) Estimate the temperature rise of the water in the tank during the day.

Solution:

$$\text{Water energy gain per day} = (\text{rate at which the system receives solar energy}) * (\text{collector area})$$
$$= \left(10.5\ \text{MJ/m}^2\right) * \left(4.8\,\text{m}^2\right)$$
$$= 50\,\text{MJ}$$
$$= 50000\,\text{kJ}$$

We use the equation $Q = mc\Delta T$.

(Q = heat gained by the system; m = mass of water; T = temperature, c_p—specific heat of heat transfer fluid (J/kg K))

Noting that density of water is 1.0 g/m^3,

$$
\begin{aligned}
m &= (\text{density of water}) * (\text{volume}) \\
&= (1.0) * (325 \times 1000)\ \text{gms} \\
&= 325\ \text{kg} \\
\Delta T &= Q/(mc) = 50000\ \text{kJ}/(325\ \text{kg} * 4.18\ \text{J/GM/°C}) \\
&= 37\text{°C}.
\end{aligned}
$$

The temperature of the water increases by 37 °C during the day.

11. Determine the payback period of a solar system costing Rs.1,000,000 and in the solar system and you save Rs.100,000 per year; the

Solution:

$$
\begin{aligned}
\text{Solar payback period} &= \text{Total investment} \div \text{savings per year.} \\
&= 1{,}000{,}000/100{,}000 = 10\text{Years.}
\end{aligned}
$$

12. **You are designing a solar car as in figure below, with a total roof area for solar cells of 6.4 m^2. Calculate the electrical power available, assuming total cell efficiency of 17% and a constant light intensity of 980 W/m^2.**

Solution:

$$
\begin{aligned}
&\text{Power input} = (\text{light intensity}) * (\text{roof area}) \\
&= (980) * (6.4) \\
&= 6.3 \times 10^3\text{W} \\
&= 6.3\ \text{kW}
\end{aligned}
$$

$$\text{Since Efficiency} = (\text{Power output})/(\text{Power input}),$$

$$\begin{aligned}\text{Power output} &= \text{Efficiency} * (\text{Power input}) \\ &= 0.17 * (6.3) = 1.06\,\text{KW}.\end{aligned}$$

13. A solar hot water system as in figure below receives solar energy at the rate of 10.5 MJ/m^2 per day. If the collector area is 4.8 m^2, collector efficiency is 0.7 and the water volume is 325 dm^3,
(a) **Calculate the total water energy gain per day;**
(b) **Estimate the temperature rise of the water in the tank during the day.**

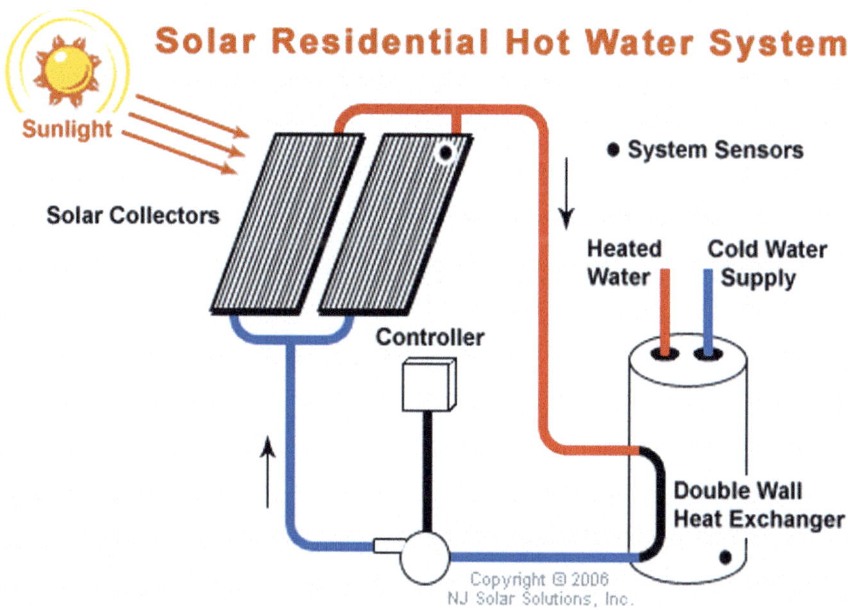

Water energy gain per day = (rate at which the system receives solar energy) * (collector area)

$$\begin{aligned} &= (10.5) * (4.8) \\ &= 50\,\text{MJ} \\ &= 50000\,\text{kJ}\end{aligned}$$

We use the equation $Q = mc\Delta T$.
(Q = heat gained by the system; m = mass of water; T = temperature, c_p—specific heat of heat transfer fluid (J/kg K)).

Noting that density of water is 1.0 kg/dm^3,

$$\begin{aligned} m &= (\text{density of water}) * (\text{volume}) \\ &= (1.0\,) * (325 \times 1000)\ \text{gms} \\ &= 325\ \text{kg} \\ \Delta T &= Q/(mc) = 50000\,\text{kJ}/(325\,\text{kg} * 4.18\,\text{J/GM/}^\circ\text{C}) \\ &= 37^\circ\text{C}. \end{aligned}$$

$$\begin{aligned} \Delta T &= Q/(mc) = 50000/(325 * 4.18) \\ &= 37^\circ\text{C}. \end{aligned}$$

The temperature of the water increases by 37 °C during the day.

14. **A single solar cell on illumination by insolation of about 800 W/m^2 produces a voltage of 0.5 V and a current up to 2.0 A. The efficiency of the solar cell is 12.5%. Find the area of the cell is.**

Solution:
Solar Insolation (SI) = 800 w/m^2

$$\text{Voltage(V)} = 0.5\,\text{V}$$

$$\text{Current(I)} = 2.0\,\text{A}$$

$$\eta = 12.5\%$$

$$\eta = (P_{out}/\ P_{in})\ \times\ 100 \tag{4}$$

$$\text{Solar Insulation(SI)} = P_{in}/\text{Area} \tag{5}$$

$$\text{Pout} = \text{Voltage(V)} \times \text{Current}(I) \tag{6}$$

Using (4), (5) and (6) we can deduce.

$$\eta = (\text{V} \times \text{I})/((\text{SI}) \times \text{Area})$$

$$\text{Area} = (\text{V}\ \times\ \text{I})/((\text{SI}) \times \eta)$$

$$(0.5 \times 2.0\,)/(800 \times 0.125)$$

$$\text{Area} = 10^{-2}\,\text{m}^2$$

Formulae:

1. **Electricity generated in output of a photovoltaic system** [2]:

$$\text{Area} = 10^{-2}\,\text{m}^2$$

where:

E Energy in kWh.
A Total Area of the panel in m^2
r % solar panel yield,
H Annual average solar radiation on tilted panels.
PR Performance ratio constant for losses (range between 0.5 and 0.9, default value = 0.75).
r Yield of the solar panel.
r Electrical power (in kWp) of one solar panel / the area of one panel.

2. **No. of Solar Panels required for an application** $= \frac{(HourlyPowerconsumptioninWattsxPeakSunlightinthearea)}{PanelWattage}$.
3. **Battery Back-Up Time = Total Load x No. of Hours backup is needed.**
4. $Backup\ Time = \frac{[Battery\ Capacity(Ah) \times Battery\ Voltage(V) \times Battery\ Efficiency(\%)]}{Connected\ Load\left(\frac{W}{h}\right)}$
5. **To determine the total solar energy received by an area a over a period of time is:**

$$E = H \times A$$

E Energy received by the collector area (J).
H Irradiation (J/m^2).
a Irradiance (W/m^2).
A collector Area (m^2).

6. **Total Power Output of a Solar system = Total Area x Solar Irradiance x Conversion Efficiency.**
7. **Solar declination is the angle between the rays of the sun and the plane of the earth's equator** [3]. It varies by an angle of up to $\pm$ 23° 27′ ($\pm$ 23.45°).

   The solar declination, in degrees $\delta = 23.45[360\text{Sin}(284 + n)/365]$.
   Where n is the day of the year.

8. **Hour Angle (ω)** [4] **= 15(t − 12) Where t = time of the day**
9. **Solar Panel Efficiency Calculation** [5]

Solar panel efficiency, in simple words, represents the amount of sunlight a solar panel can convert into usable electricity. It is calculated by dividing the amount of electricity produced by the solar panel by the amount of the sun that strikes the solar panel.

Formula:

Here is the mathematical representation of the solar panel efficiency.

$$\text{Efficiency}(\%) = [(\text{Pmax} \div \text{Area}) \div 1000] \times 100\%.$$

where

Pmax	Maximum solar panel power (measured in watts).
Area	Length × Width of solar panels (measured in sq. M).
1000	Conversion factor to convert power output per unit area from watts per sq. M to percent.

10. **Solar payback period = Total investment ÷ savings per year.**

8.3 Wind Energy [6]

Solved Problems:

1. **Wind at a speed of 10 m per second passed through a wind turbine with a blade radius of 30 m. What is the theoretical power available and the maximum fraction of power available?**

   SOL:

$$\text{Kinetic Energy/Second} = \text{Power} = 0.5 \times \rho \times \text{A} \times \text{V}^3 (\text{Watts})$$

$$= 0.5 \times 1.2256 \times 3.146 \times 30 \times 30 \times 10 \times 10 \times 10 = 1,735,081$$
$$\text{W} = 1735\,\text{kW} = 1.735\,\text{MW}$$

is the maximum theoretical power for these conditions

Applying Betz Law the maximum fraction of this which can be captured is = 1.735 × 0.593 = 1.03 MW.

2. **For example at 10 m/s the turbine has an output of 1000 kW and the wind blows at 10 m/s for 500 h per annum. Find the electricity produced.**

$$\text{Electricity Production at } 10\text{m/s} = 1000\,\text{kW} \times 500\,\text{hrs}$$

$$= 500,000\,\text{kW h}$$

3. Determine the power in the wind if the wind speed is 15 m/s and blade length is 50 m.

Solution:
Given:

Wind speed $V = 15\,m/s$,
Blade length $l = 50\,m$,
Air density $\rho = 1.23\,kg/m$.

The area is given by,

$$A = \pi \times 50 \times 50$$
$$= 7857\,m^2$$

The wind power formula is given as, $P = 1/2 \times \rho \times A \times V^3$.

$$= 1/2 \times 1.23 \times 7857 \times 15^3$$

$$P = 16308\,MW$$

4. **A wind turbine travels with the speed of 20 m/s and has a blade length of 20 m. Determine wind power.**

Solution:
Given:

Wind speed $V = 20\,m/s$,
Blade length $l = 20\,m$,
Air density $\rho = 1.23\,kg/m^3$,
Area $A = \pi \times 400 = 1257\,m^2$.

The wind power formula is given as,

$$P = 1/2 \times 1.23 \times 1257 \times 20^3$$

$$P = 6184.44\,MW$$

5. **Calculate the annual electricity production of a wind turbine, give that 10 wind turbines each of 1.75 MW are the be installed, swept area is 3142m^2. Consider mean annual wind speed 8 m/s [k = 3.2].**

Given:

V mean annual wind speed 8 m/s.
T No. of wind turbines to be installed.

A_t swept area is 3421 m^2.

$$\text{Annual Wind Electricity Production} = K \times V^3 \times A_t \times T$$

$$= 3.2 \times 8 \times 8 \times 8 \times 3142 \times 10 = 56,049,664\,\text{kWh} = 56\,\text{GWh}$$

6. The following data relate to a wind turbine:

Velocity of wind at 15°C = 10 m/s.
Turbine diameter = 10 m.
Operating speed of the machine = 35 rpm at maximum efficiency of 40%
Calculate:

(i) Total power density in the wind stream
(ii) The maximum power density
(iii) The actual power density
(iv) Power output of the turbine.

Wind Formulae:

1. The wind energy formula is given by,

$$P = \frac{1}{2}\rho A V^3$$

where,

P power,
ρ air density,
A swept area of blades given by.
V velocity of the wind m/s.

1. Electricity Production at particular wind speed = Power at particular wind speed from Power Curve × No. of Hours of particular wind speed from Wind Speed.

$$\text{Annual Electricity Production(kWh)} = K \times V^3 \times A_t \times T\text{where}$$

K 3.2 (approximation factor).
V Mean Annual Wind Speed (m/s).
A_t Swept Area of wind turbine (m^2).
T Number of Wind.

Turbines.

This equation must be treated with caution as the factor, K, is used to approximate the relationship between wind speed and distribution. This relationship will change depending on the particular site. The equation does not allow for optimization based on particular turbine power curves either.

8.4 OCEAN Energy [7]

1. **The basin area of a tidal power plant is 20×10^6 m². The tidal range is 8 m, calculate the energy generated in KWh.**

The energy available from barrage is dependent on the volume of water.
The potential energy contained in a volume of water is:

$$E = 1/2A\rho gh^2$$

where:

h is the vertical tidal range,
A is the horizontal area of the barrage basin,
ρ is the density of water = 1025 kg per cubic meter (seawater varies between 1021 and 1030 kg per cubic meter) and
g is the acceleration due to the Earth's gravity = 9.81 m per second squared.

$$E = 1/2A\rho gh^2$$
$$= 1/2 * 20 \times 10^6 * 1025 * 9.81 * 8 * 8 = 6.4\,\text{MW}$$

8.5 Additional Problems

8.5.1 *Problems on Renewable Energy Power Plants*

1. A small household-lighting system is powered from a nominally 8 V (i.e. four cells at 2 V) storage battery having a 30 Ah supply when charged. The lighting is used for 4.0 h each night at 3.0 A. Design a suitable photovoltaic power system that will charge the battery from an arrangement of Si solar cells.
(a) How will you arrange the cells?
(b) How will the circuit be connected?
(c) How will you test the circuit and performance?

Assume: Cells should produce about 10 V to charge an 8 V storage battery.

2. Calculate the approximate time for a single-crystalline PV module to generate electricity equal in energy terms the primary energy used in its manufacture. Consider a typical Japanese climate with 1450 kWh (m^2 y^{-1}) of insolation, modules of 15% efficiency and 1.0 kWp peak power of modules. The total direct manufacturing and processing energy (includes refining, etc.) is 1350 kWh/kWp,

plus a further 1350 kWh/kWp of factory 'overhead energy', with 90% of both from electricity (generated thermally at 55% efficiency from fuels in combined cycle generation).

3. A number of designs of wind turbine pass the output wind from one set of blades immediately on to a second identical set (e.g. two contrary rotating blades set on the same horizontal axis). By considering two actuator disks in series, and using linear momentum theory, show that the combined maximum power coefficient CP equals 0.64.
4. **Battery Sizing for an Off-Grid Cabin**. A cabin near Salt Lake City, Utah, has an AC demand of 3000 Wh/day in the winter months. A decision has been made to size the batteries such that a 95% system availability will be provided, and a back-up generator will be kept in reserve to cover the other 5%. The batteries will be kept in a ventilated shed whose temperature may reach as low as – 10 °C. The system voltage is to be 24 V, and an inverter with overall efficiency of 85% will be used.

Assume: we need about 4.6 days of storage.

5. A 40-m, three bladed wind turbine produces 600 kW at a windspeed of 14 m/s. Air density is the standard 1.225 kg/m^3. Under these conditions,
   a. At what rpm does the rotor turn when it operates with a TSR of 4.0?
   b. What is the tip speed of the rotor?
   c. If the generator needs to turn at 1800 rpm, what gear ratio is needed to match the rotor speed to the generator speed?
   d. What is the efficiency of the complete wind turbine (blades, gear box, generator) under these conditions?
6. **A Loan to Pay for a Small Wind Turbine**. Suppose that a 900-W Whisper H900 wind turbine with 7-ft diameter (2.13 m) blade costs $1600. By the time the system is installed and operational, it costs a total of $2500, which is to be paid for with a 15-yr, 7 percent loan. Assuming O&M costs of $100/yr, estimate the cost per kWhr over the 15-year period if average wind speed at hub height is 15 mph (6.7 m/s).
7. Design the rotor radius for a multi-blade wind turbine that operates at a wind speed of 20 kmph to pump water at a rate of 4 m^3 /hr with a lift of 10 m. Also calculate the angular velocity of the rotor in rad/s.

Given: water density = 1000 kg/ m^3, g = 9.8 m s^2, water pump efficiency = 60%, efficiency of rotor to pump = 75%, Cp = 0.3, $\lambda = 2.0$, air density = 1.2 kg/m^3.

References

1. Environmental science: numerical questions—solar cell: calculate efficiency and area, 2023
2. https://www.saurenergy.com/solar-energy-blog/here-is-how-you-can-calculate-the-annual-solar-energy-output-of-a-photovoltaicsystem
3. https://www.myelectrical2015.com/2024/02/concept-of-solar-declination-angle.html

4. https://arka360.com/ros/solar-angles/
5. https://www.photonicuniverse.com/en/resources/articles/full/7.html
6. https://www.e-education.psu.edu/emsc297/node/649
7. https://brainly.in/question/31355675

Chapter 9
Measurement for Enhancing Quality and Efficiency of Renewable Energy Systems

9.1 Introduction

The quality and efficiency of any system can be maintained or further improved, only when we are able to measure the various parameters of the system, compare them with the standard value and accordingly adjust the operating parameters. The advent of modern technologies has enabled develop several modern measuring instruments [1] which are not only digital but also computerized and enabled by artificial intelligence and IoT techniques to give us accurate and reliable data and make computations faster.

This chapter presents several modern day instruments related to measuring of various parameters of renewable energy systems.

9.2 Need for Parameter Measurement

Measurement offers several benefits to Hybrid Power System (HPS) operations as:

- Optimize generation, transmission and distribution of power.
- Balance Supply and Demand.
- Reduce losses.
- Increase efficiency.
- Safety.
- Improving co-ordination of protection devices as relays and circuit breakers.
- Reduced costs.
- Better maintenance of equipment to increase life-time of equipment's as transformers, relays, circuit breakers.

N. Manohar et al., *Machine Learning Applications in Renewable Energy*, Green Energy and Technology, https://doi.org/10.1007/978-981-97-9939-8_9

9.3 Electrical Parameters Measured

The electrical parameters measured by various measuring instruments are [2, 3]:

- Voltage Magnitude—instantaneous and continuous,
- Voltage Phase Angle—instantaneous and continuous,
- Current—instantaneous and continuous,
- Current Phase Angle—instantaneous and continuous,
- Network frequency,
- Phase Angle,
- Active Power,
- Reactive Power,
- Sags and swells in voltage values,
- Power factor,
- Harmonic currents,
- Power,
- Energy,
- Inrush current values,
- Light flicker,
- Rotational speed of generators,
- Torque,
- Battery charge levels,
- Solar irradiance,
- Wind Speed.

9.4 Solar System Efficiency

The efficiency of a photovoltaic (PV) cell, or solar cell, is the percentage of the solar energy falling on a PV device that is converted into usable electricity [4]. The High efficiency is obtained by exposing the cell to a constant, standard level of light while maintaining a constant cell temperature, and measuring the current and voltage that are produced for different load resistances.

The various factors that affect a cell's efficiency are:

i. **Location dependant factors**: The dispersion and intensity of solar radiation, is different at different locations.
ii. **Temperature**—As the temperature increases the efficiency of the solar cell decreases, that is efficiency is inversely proportional to the temperature.
iii. Load on the system
iv. Dust
v. Solar Panel material.

The Maximum solar cell efficiency is given by the following Eq. (9.1) [5].

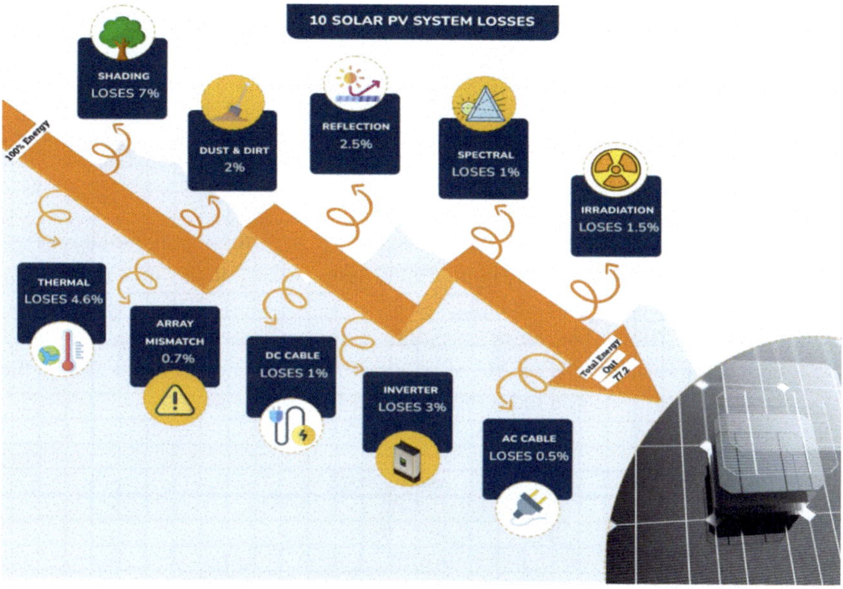

Fig. 9.1 Losses in solar system

$$\begin{aligned}
&\text{Solar cell Efficiency (Maximum)}\\
&\eta_{\max} = \frac{P_{\max}}{E * A_c} \times 100\%\\
&P_{\max} = \text{Maximum Power Output (in W)}\\
&\text{E} = \text{incident radiation flux (in W/m}^2\text{)}\\
&A_c = \text{Area of Collector (in m}^2\text{)}
\end{aligned} \quad (9.1)$$

Also, we know that for any system:

$$\text{Power Output} = \text{Power Input} - \text{Losses} \quad (9.2)$$

Therefore to improve efficiency, we need to know the various losses and the methods to reduce them.

The various losses in a solar system [6] Fig. 9.1 are.

9.4.1 *Solar PV Losses*

1. **Shading Effects on Solar Panel System**

Shading directly impacts solar panels' performance [7]. Shading reduces the efficiency of photovoltaic systems because it blocks sunlight from directly reaching the solar panels.

Another type of shading is "Soft Shading." This type refers to the shadow made by the clouds as seen in Fig. 9.2. This issue may occur in some geographic locations where there are clouds. Soft shading generates diffused light. However, this issue is not as serious as the shading by object such as tree in Fig. 9.2, buildings, electricity poles, birds, antennae satellites etc.

1. **Solutions for Solar Shading Issue**

   i. **Adding By-Pass Diodes**: By-pass diodes are fitted with solar energy systems, to make current flow around those solar panels cells which may not work due to under-shading.

   ii. **Built-in Micro Inverters in Solar Energy System**: A micro-inverter converts variable DC to AC of required frequency. The output AC of the micro-inverter can be fed o the electrical load or to the grid.

Fig. 9.2 Shading due to trees and clouds

iii. **Addition of Solar Optimizers**: Solar panel optimizers use Maximum Power Point Tracking (MPPT) technology. This increases the efficiency of the panels. Solar optimizers are Module Level Power Electronics (MLPE) devices attached to solar panels during installation. They ensure that each panel generates optimal energy output. The MLPE can be installed to one or every solar panel in the array

2. **Dust Effects on Solar Performance**

Dust particles also impact the performance of the solar photovoltaic system. While making a solar design and calculations, energy experts should consider this energy system loss also. Solar panel arrays need to be cleaned once every six months. This not only increases energy output by 3 to 5% on average but also improves efficiency by up to 25% in really dusty places.

3. **Solar Cell Array Optical Losses**

Optical losses happen when light bounces off the panel's surface.

4. **Solar Cells Spectral Responses**

This system loss is about the output of different types of wavelengths that solar cells receive. Solar energy cells do not use all sun's wavelengths. There are different frequencies, but the most common ones are 4% UV, 54% infrared, and 43% visible light. To enhance electrical output, manufacturers work to design panels so that they offer a generic, broad response.

5. **Solar Energy System Loss Due to Irradiance Level**

System loss due to irradiance level shows the reduction in solar efficiency when the irradiance level decreases from the STC rating (Standard test Conditions) of 1000W/m^2 to a low irradiance of 200W/m^2.

6. **Solar Panel Cells Thermal Loss**

This loss is due to an intrinsic feature of the solar cell structure.

7. **Solar Panels DC Wiring Losses**

Losses in DC connections cannot be eliminated because whenever a current flows, energy is lost.

8. **Losses in Solar Power Inverters**

The efficiency of the most popular type of inverter, the string inverter, is 97% Temperature and load are two of the key variables impacting solar inverter efficiency. If the unit is positioned outdoors or inside, make sure it is well-ventilated and sheltered from the sun's glare. The efficiency curve of inverters is similar to that of most devices with maximum efficiency occurring around the maximum working load. This is typically not a problem until the inverter load goes below 25–30%, but below this point, it drops sharply.

Fig. 9.3 Thermal imaging camera

9. **Energy System Voltage Drop**

As the current flows into the electrical circuit, a voltage occurs due to the resistance of the wires. This principle, known as a voltage drop, leads to a minor loss of output from your solar array. A longer wiring run adds more resistance to the circuit which leads to a higher voltage drop. It is generally considered best practice to keep the voltage down to 3% or less, although many devices fall far below that level.

Voltage drop has a straight effect on the production of the solar system. If the wiring run is too long, your panels can not supply the inverter with enough voltage. The quality of the entire solar system will suffer.

9.5 Measuring Instruments for Solar Energy [8]

9.5.1 Thermal Imager or Thermal Imaging Camera

Temperature measurements are critical to the solar system. Thermal imaging cameras (Fig. 9.3) can detect hot spots in electrical and circuit breaker panels or on photovoltaic modules. They can monitor the level of cooling fluids, such as dielectric oil in transformers.

9.5.2 Digital Multimeter

A digital multimeter (Fig. 9.4) is a measuring instrument that can measure quantities such as voltage, current, and resistance.

Fig. 9.4 Digital multimeter

9.5.3 Clamp Meter

A clamp meter Fig. 9.5 is an electrical testing tool that has a basic digital multimeter with a current sensor. Clamps measure current. Probes measure voltage. The hinged jaw securely clamps around a wire, cable, or conductor, facilitating current measurement in the circuit without the need for disconnection or de-energization.

9.5.4 Solar Power Meter

Solar Power Meters Fig. 9.6 measure and monitor the sunlight intensity, overall energy, efficiency and placement of all kinds of Solar Systems. A Solar Power Meter helps determine the amount of sunlight or the amount of solar energy received by a given area in W/m^2 which is known as solar constant. This helps immensely in determining optimum placement of solar panels.

Digital Insulation Testers measuring insulation resistance up to 10 kV are suitable for Solar power systems, as they have to test high voltage equipment like cables, switchgear, motors and conductors. Insulation Testers can be used in a variety of tests including identifying power quality failures, breakdown tests, spot checks as well as timed tests.

Fig. 9.5 Clamp meter

Fig. 9.6 Solar power meters 9.3.5 insulation tester

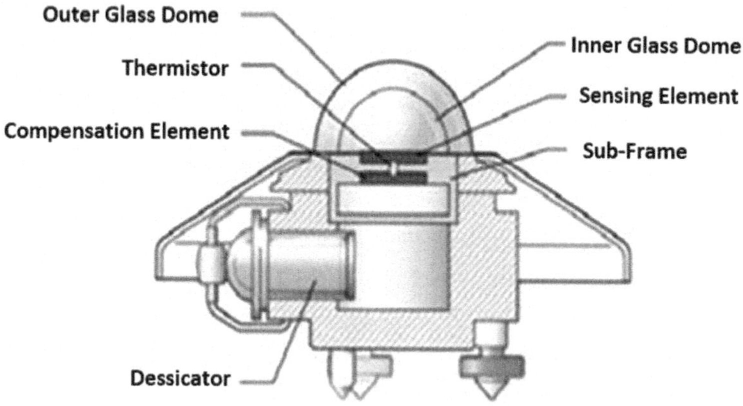

Fig. 9.7 Pyranometers

9.5.5 *Pyranometers*

Pyranometers Fig. 9.7 are radiometers designed for measuring the irradiance on a plane surface, normally from solar radiation and lamps.

9.5.6 *Pyrheliometers*

A pyrheliometer Fig. 9.8 is a device that measures solar irradiance coming directly from the sun. The SI units of irradiance are watts per square metre (W/m^2).

Fig. 9.8 Pyrheliometer

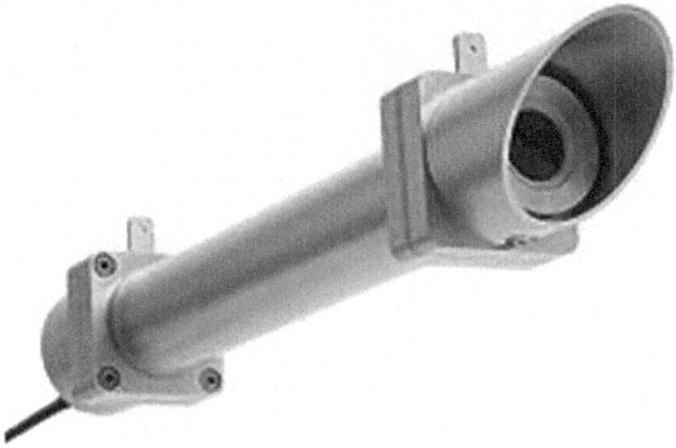

9.5.7 Sun Trackers

Solar tracker Fig. 9.9 or solar tracking system adjusts the face of the solar panel or reflective surfaces to follow the movement of the Sun. The movement of solar trackers increases the solar energy output by up to 40% than standard panels.

9.6 Measuring Instruments for Grid [9]

Analysis of measurement and initiation of corrective action can be done only when the parameters can be sensed. Therefore, we shall also study about several sensors in the grid apart from measurement and automatic computational devices. Namratha in her chapter "Sensing Layer—Layered Architecture for Efficient Operation of Intelligent Power System", has explained the role, types, functionality and importance of sensors in the present day Intelligent Power Systems (IPS).

Sensors are classified based on several criterion as:

i. **Based on input/output signals**: Analog or Digital.
ii. **Based on the principle**: Chemical, Magnetic, acoustic, pressure, acceleration, vibration
iii. **Based on Parameter**: Current, voltage, Temperature, humidity.

The classification is not mutually exclusive, meaning a temperature sensor may be of analog or digital. The various types of sensors placed at various locations in a power grid are indicated in Fig. 9.10, and placement on one of the most important device of a power system which is a transformer can be seen in Fig. 9.11

Dual Axis Solar Tracker

Fig. 9.9 Solar tracker

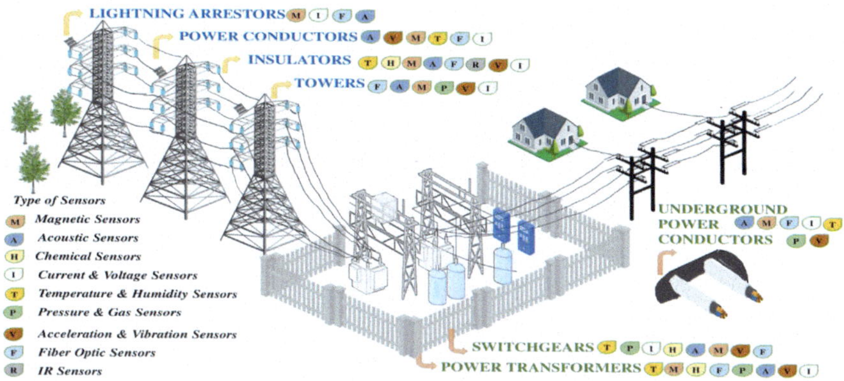

Fig. 9.10 Various sensors and their locations in a power grid

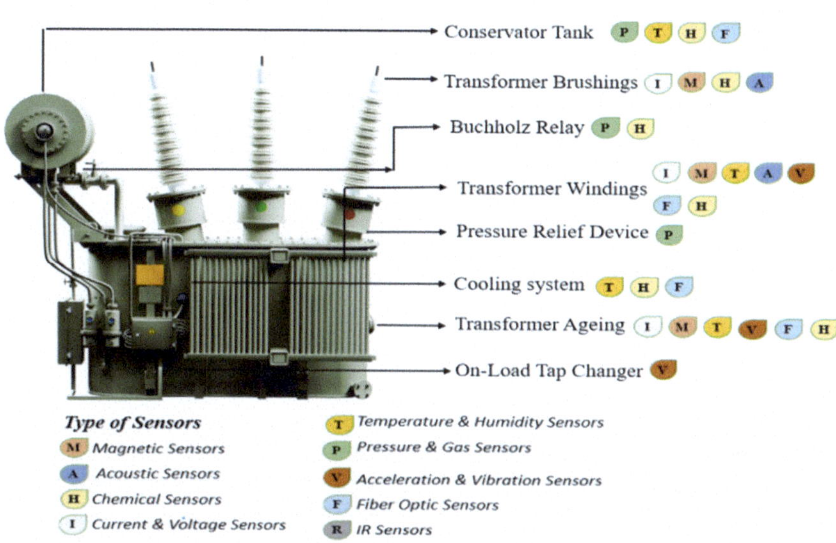

Fig. 9.11 Various sensors on a transformer

9.6.1 Line Sensors

Line sensors Fig. 9.12 on the section of the transmission line help minimize the impact of the outage. New line sensors are node monitoring systems known as a fault indicator that detects and sends an alert when problems occur so that line crews can service faulted equipment with minimal delay.

Line sensor deployments at leading U.S. utilities have demonstrated significant improvements in system average interruption duration index (SAIDI), system average

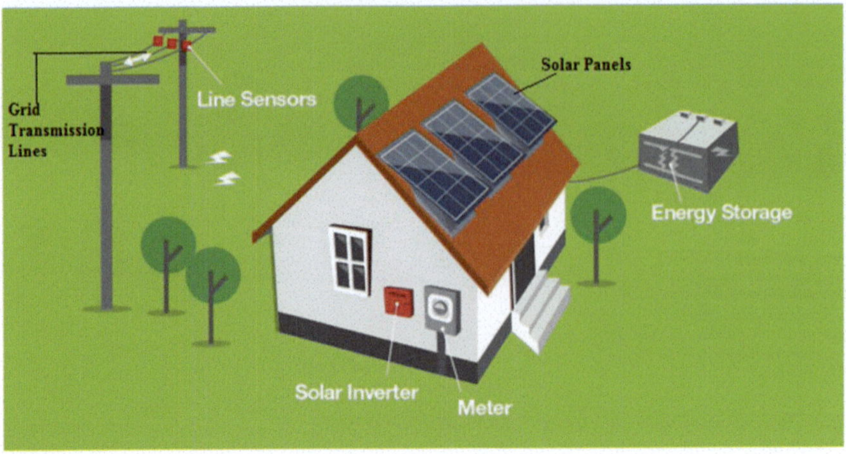

Fig. 9.12 Line sensors

interruption frequency index (SAIFI) and customer average interruption duration index (CAIDI).

9.6.2 Three-Phase Power Quality Analyzers

Three-phase power quality analyzer Fig. 9.13 help you locate, predict, prevent and troubleshoot problems in three- and single-phase power distribution systems. Troubleshooting is faster with on-screen display of trends and captured events, even while background recording continues.

Models as AIV.1 Fluke Power Quality Analyzer, by AAA Testers have additional features Utilizing as Energy Loss Calculator function. It computes and displays the fiscal cost of energy wasted due to poor power quality. This energy monetization capability identifies the most energy-wasteful areas.

https://www.mdpi.com/1996-1073/15/19/7339

9.6.3 Digital Fault Recorders

A digital fault recorder (DFR) is an intelligent electronic device (IED) which samples binary data during power system transients, using communications to retrieve fault, disturbance and sequence of event records, captured by protection relays. Harmonics, frequency and voltage levels are some examples of data captured by DFRs. They enable local substation troubleshooting and data archiving for permanent storage and analysis.

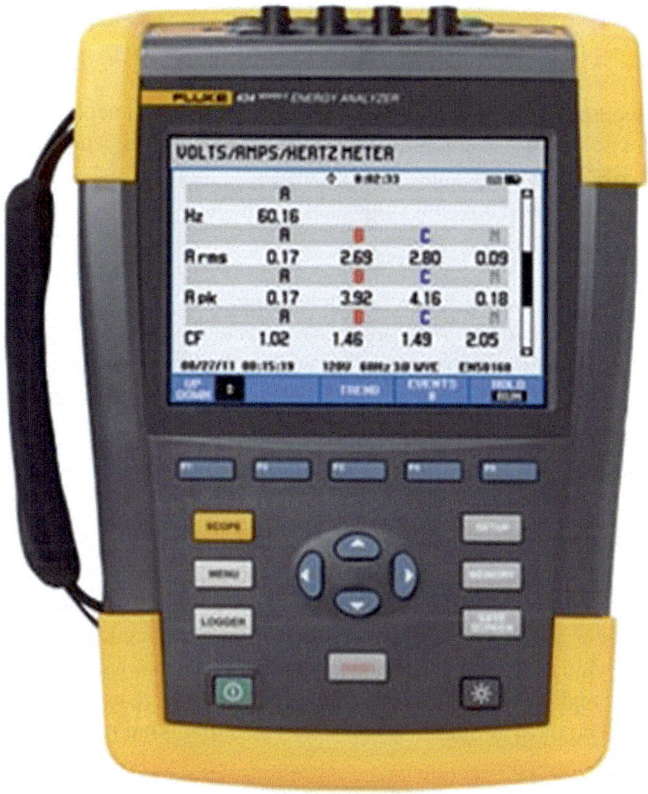

Fig. 9.13 Three-phase power quality analyzer

DFRs have generally three types of recording mechanism:

- High-speed disturbance recording of the instantaneous waveform signals of both current and voltages.
- Low-speed disturbance recording that is used to capture both long-term and short-term disturbances. In this recording type, the DFR records calculated parameters at a relative high sampling rate (1 – 10 waveform cycles).
- Steady-state recording that is used to capture min, max and average values of calculated parameters such harmonics at relatively low time resolution.

Two recording triggering methods are commonly used:

- Events that the DFR calculates from its input signals.
- Digital signals triggered by the protection equipment.

DFRs:

- Provide a permanent detailed record of all substation activity.
- Record Transfer Trip & Block Signals and other messages.
- Provide secure data collection while also isolating the IED network.
- Permanent recording of internal protection relay operands and calculations in Sequence of Events (SOE) and Fault records.
- Have a fairly long lifetime and simplified maintenance.

DFRs have been extensively used:

- For substation distributed digital fault recording.
- For monitoring of system protection performance.
- As components in larger enterprise wide fault and disturbance recording systems.
- Monitoring of transient events, e.g. in the case of field experiments, connecting new assets to the grid.

9.7 Measuring Instruments for Wind Turbines

Wind resource measurement is an important [10] aspect of wind power development. Information about how hard the wind blows and in what directions determines how much power a wind turbine would produce. A variety of technologies are available to measure wind conditions. The two most important characteristics of wind are speed and direction. The most common systems for collecting wind resource information are listed below.

The various sensors and measuring instruments [11] and their location on the wind turbine are shown if Fig. 9.14. We shall study some of the measuring instruments important for wind turbine operation and maintenance. The common measuring devices as thermometer and pressure meters will not be delved into. Most of these devices have now become IoT based in that the value of the parameter is sent to a computing device and control signals are generated automatically from remote locations.

9.7.1 Anemometers

Anemometers are conventionally used for wind speed measurement. The classification of anemometers in given in Fig. 9.15.

The principle of the various anemometers and their figures is presented in Table 9.1.

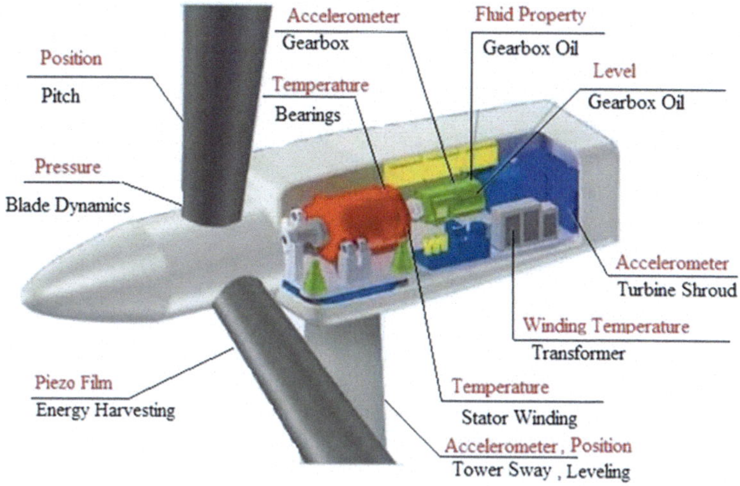

Fig. 9.14 Location of measuring instruments on wind turbine

Fig. 9.15 Classification of anemometers

Anemometers

Velocity anemometers

1. Cup anemometers
2. Vane anemometers
3. Hot-wire anemometers
4. Laser Doppler anemometers
5. Acoustic resonance anemometers
6. Ping-pong ball anemometers

Pressure anemometers

1. Plate anemometers
2. Tube anemometers
3. Pitot tube static anemometers

9.7.2 Meteorological Towers

Meteorological towers Fig. 9.16a are commonly used to measure wind speed and direction both on land and over water. The towers are generally between 50 m and 120 m high and as seen in Fig. 9.16b are fitted with a variety of instruments, including anemometers and weather vanes, to measure the wind speed and direction, generic sensors, data logger, power and communications system.

9.7.3 Remote Sensing

Remote sensing is a relatively new technology for measuring the wind speed. Wind speed can be measured remotely using instruments on the ground and mounted

Table 9.1 Principle of the various anemometers and their figures

Type of anemometer	Principle	Figure
1. Cup anemometers	It consisted of four hemispherical cups on horizontal arms mounted on a vertical shaft. The air flow past the cups in any horizontal direction turns the shaft at a rate roughly proportional to the wind's speed. Therefore, counting the shaft's revolutions over a set time interval give the average wind speed	
2. Vane anemometers	A *vane anemometer* combines a propeller and a tail on the same axis to obtain accurate and precise wind speed and direction measurements from the same instrument. The speed of the fan is measured by a rev counter and converted to a wind speed by an electronic chip. Hence, volumetric flow rate may be calculated if the cross-sectional area is known	
3. Hot-wire anemometers	Hot wire anemometers use a fine wire electrically heated to some temperature above the ambient. Air flowing past the wire cools the wire. As the electrical resistance of most metals is dependent upon the temperature of the metal a relationship can be obtained between the resistance of the wire and the speed of the air	
4. Laser Doppler anemometers	Laser Doppler anemometers use a beam of light from a laser that is divided into two beams, with one propagated out of the anemometer. Particulates flowing along with air molecules near where the beam exits reflect, or backscatter, the light back into a detector, where it is measured relative to the original laser beam. When the particles are in great motion, they produce a Doppler shift for measuring wind speed in the laser light, which is used to calculate the speed of the particles, and therefore the air around the anemometer	

(continued)

Table 9.1 (continued)

Type of anemometer	Principle	Figure
5. Acoustic resonance anemometers	Built into the cavity is an array of ultrasonic transducers, which are used to create the separate standing-wave patterns at ultrasonic frequencies. As wind passes through the cavity, a change in the wave's property occurs. By measuring the amount of phase shift in the received signals by each transducer, and then by mathematically processing the data, the sensor is able to provide an accurate horizontal measurement of wind speed and direction	
6. Ping-pong ball anemometers	As wind displaces the ping pong ball, the angle measured from the built-in protractor can be converted into wind speed using the conversion table	

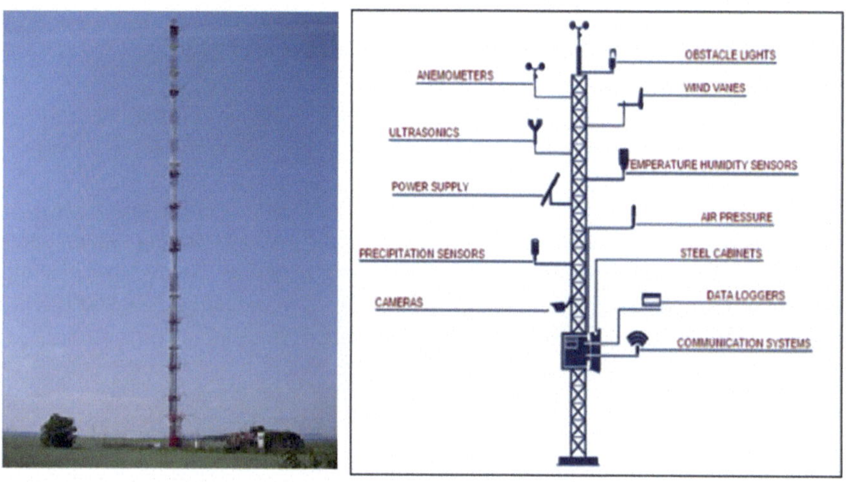

(a) (b)

Fig. 9.16 A meteorological tower **a** on field, **b** with various sensors

on airplanes or satellites. Sensors on the ground use Light Detection and Ranging (LiDAR) technology to measure the wind speed and direction. LiDAR systems send out a signal which bounces off of particles carried by the wind and returns to the sensor. The system measures the properties of the return signal to calculate the wind speed and direction. Figure 9.17 shows the placement of LIDAR on wind turbine.

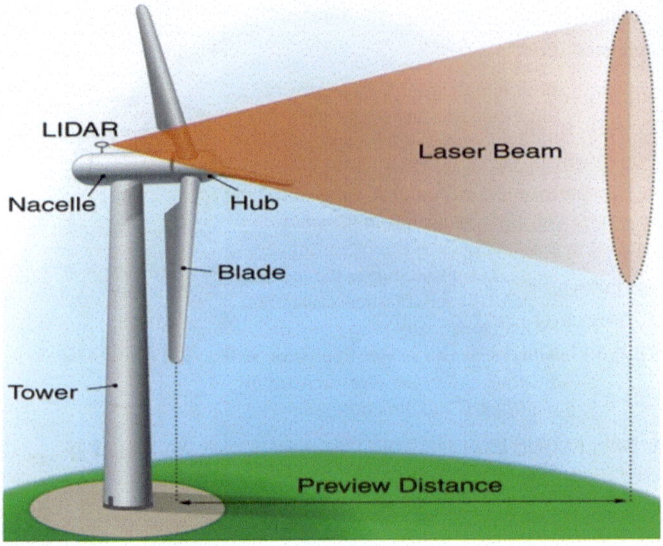

Fig. 9.17 LIDAR on wind turbine

9.8 Measuring Instruments for Waste-To-Energy Conversion Techniques [12]

9.8.1 Measurement Parameters

Measuring instruments enable monitor and control the waste-to-energy system operation. The gauges, sensors[13], switches and transmitters must be able to withstand high pressures, upto 500psi and extreme temperatures in the range 500–1600 °F.

The parameters to be measured in waste-to-energy plants are:

- Flow rate
- Pressure
- Liquid level
- Temperature
- pH
- Conductivity
- Reactivity
- Main turbine control
- Condenser level control
- De-aerator level control
- Activated carbon supply control
- Dust collector inlet temperature control
- Slaked lime feed control

- Air temperature control for preventing white smoke
- Flue-gas measurement (NO_X, SO_2, CO_2, O_2, dust).

9.8.2 Oxygen Analyzer

An oxygen analyzer is used to measure trace or percent levels of oxygen concentration in gas streams or the surrounding atmosphere. The measurement methods of the oxygen analyzers are classified into the following four categories.

1. Zirconia type measurement system
2. Paramagnetic type
3. Optical type
4. Electrochemical type.

Selection of the type of oxygen analyzer depends upon the application and usage. They are used for energy conservation, air pollution prevention, safety management, waste-to-energy systems and quality control.

We shall study the operation of Zirconia type measurement system in Fig. 9.17.

9.8.3 Zirconia Type Measurement System [14]

Concentration cell system consists of a solid electrolyte like zirconia. As shown in the Fig. 9.18, when porous platinum electrodes are attached to both sides of the zirconia element to be heated up and gases of different partial oxygen concentrations are brought into contact with the respective surfaces of the zirconia, the device acts as an oxygen concentration cell. This phenomenon causes an electromotive force (E in Volts) to be generated between both electrodes according to Nernst's equation which is proportional oxygen concentration (DO in milligrams/Litre). Equation 9.2 is the relationship between electromotive force (E in Volts) to be generated between both electrodes and oxygen concentration.

$$\mathrm{E}\,\alpha\,\mathrm{DO} \tag{9.3}$$

9.8.4 Flow Meter

Accurate measurement of landfill gas flow throughout the system is needed, whether the objective is landfill gas to energy projects, monitoring the process for optimization, or GHG emissions reporting.

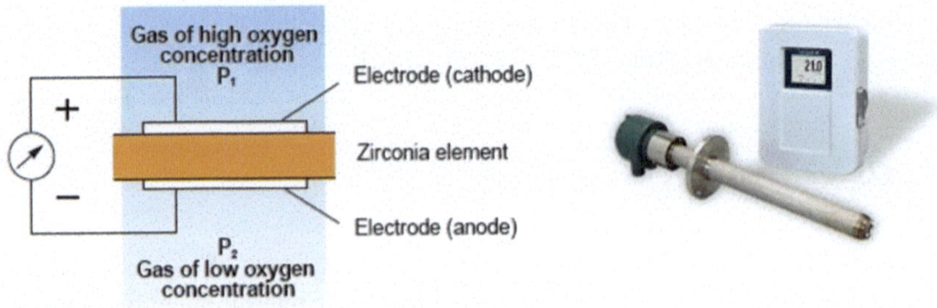

(a) Zirconia Oxygen Analyzer Principle of Operation (b) Zirconia Oxygen Analyzer Instrument

Fig. 9.18 Zirconia type oxygen analyzer

Magnetic Flow Meter

- According to Faraday's electromagnetic induction law, electromotive force is induced in a conductor when moving through a magnetic field. A magnetic flow meter uses this law to measure the flow rate by considering a conducting fluid as the conductor and providing electrodes to transmit the induced electromotive force externally.
- The ADMAG magnetic flow meter adopts the dual frequency excitation method to suppress fluid noise and achieve high stability. It accurately measures the flow rate of electrical conducting fluids such as purified water and waste water, thus helping to reduce energy consumption as well as prevent global warming.
- The ADMAG AXR series two-wire magnetic flow meter, launched in 2009, achieves an significant power saving of 96–99% compared with our existing four-wire magnetic flow meter. Moreover, it doesn't require any AC power and can apply a two-wire system, dramatically reducing the initial instrumentation cost.

Turbidimeter

- Yokogawa offers turbidimeters based on different measurement methods suitable for various conditions. The laser turbidimeter is ideal for detecting membrane ruptures or monitoring status during filtration through membrane, which is an advanced processing method for purifying water. The right angle scattered light turbidimeter identifies turbidity by measuring transmitted light, because light incidence into water is reflected and scattered on particles, and the amount of light transmitted decreases in proportion to the length of optical path and turbidity.
- Yokogawa's turbidimeters accurately measure the turbidity of whichever purified or wastewater, thus contributing to the reliance and safety of social infrastructure and environmental preservation (Fig. 9.19).

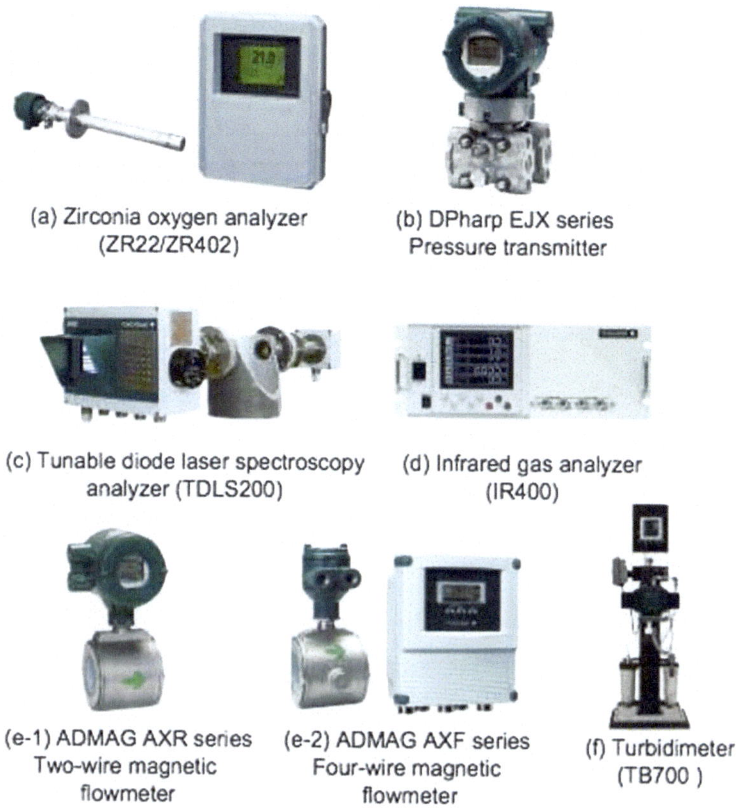

Fig. 9.19 Major industrial measuring instruments

References

1. https://www.studyforfe.com/blog/power-measurement-and-instrumentation/
2. https://www.mermecgroup.com/pageview2.php?i=239&sl=1
3. https://instrumentationtools.com/measurement-of-electrical-parameters-in-steam-turbine/
4. https://solargroup.co.nz/solar-panel-efficiency/
5. https://www.cleanenergyreviews.info/blog/most-efficient-solar-panels
6. https://solarsme.com/top-solar-energy-system-losses/
7. https://ratedpower.com/blog/shading-losses/
8. https://www.metravi.com/list-of-essential-test-measuring-instruments-for-solar-technicalprofessionals/?srsltid=AfmBOooazVIiMr_almoz7wWUSqFKFk5AhLZvrRdqx57lqFgqsivgk5f5
9. https://www.wiha.com/int/en/service/knowledge/electrical-measuring-devices/
10. https://cdsentec.com/exploring-the-role-of-measuring-instruments-in-the-wind-power-generation-process/
11. Karad S, Thakur R, Efficient monitoring and control of wind energy control systems using Internet of Things (IoT): a comprehensive review 23. https://doi.org/10.1007/s10668-021-01267-6
12. https://blog.wika.us/applications/robust-measuring-solutions-for-municipal-energy-recovery-facilities/

13. Namratha Manohar J, Sensing layer—layered architecture for efficient operation of intelligent power system. In: Emerging IT/ICT and AI technologies affecting society. https://doi.org/10.1007/978-981-19-2940-3_1
14. https://www.yokogawa.com/library/resources/yokogawa-technical-reports/yokogawas-industrial-measuring-instruments-for-energy-conservation-and-environmental-preservation/

Chapter 10
Energy from Bio Mass–Bio Gas Plants

10.1 Introduction

Bio mass is a natural existing element. It is consumed by the living organisms of the universe. As is known every system has an input and output. Therefore, if bio mass is one of the input for a living organism system, then after digestion and the living body consuming the useful elements as vitamins, proteins and other minerals the un-digested biomass output of the system is the biomass[1]. So, it is understood that as long as the universe exist there will be bio mass waste. Bio mass waste is also generated from the natural sources as the leaves after drying, flowers and fruits as the fall and are not consumed etc.

With the depletion of natural sources of energy, there is need to research for alternate sources of energy. One, prominent, easy and perpetually available waste is bio mass. Hence, developing modern techniques for converting the biomass waste into various forms of energy is the need of the hour. The various techniques have been detailed in Chapter 2: Waste to Energy. In this chapter the various Bio Gas Plants and their design criteria is presented.

10.2 Bio Gas Plants

10.2.1 Principle of Operation of Bio Gas Plants

Bio gas is obtained from Bio mass. Biomass is burned to convert to renewable liquid, and gaseous fuels through various processes. Anaerobic digestion is one such process, in which organic matter, such as food or animal waste, is broken down by microorganisms in the absence of oxygen. Waste is enclosed in enclosed environment called bio gas plant commonly known as gobar gas plant in India. Gobar in some

N. Manohar et al., *Machine Learning Applications in Renewable Energy*, Green Energy and Technology, https://doi.org/10.1007/978-981-97-9939-8_10

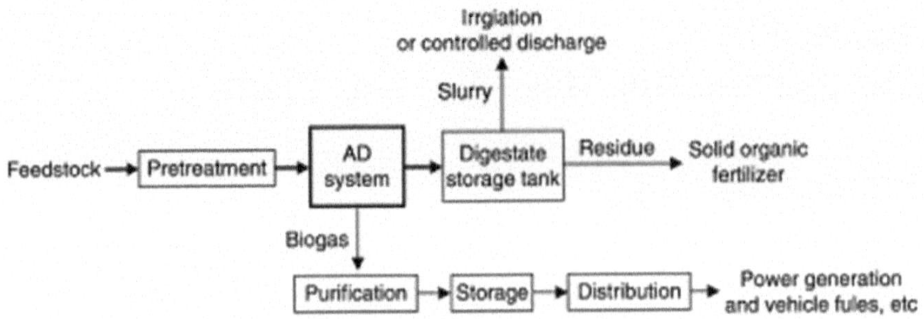

Fig. 10.1 Process of Bio Gas Plant

Indian languages means cow dung, which is the input for the gobar gas plant, as it is abundant in India.

Bio mass after pre-treatment is mixed with the required proportion of water to make a slurry. The slurry is kept in a tank called a digester. The number of days kept in the digester is called 'Retention Time'. Retention Time depends upon the type and quality of the input bio mass. During the period when the slurry is held in the digester, anaerobic digestion (AD) takes place. The methane gas produced as the output can be used for cooking, generation of electricity [2], and some other applications as production of methanol. The digested residue after the digestion period, which is fully digested will form an organic fertilizer. Figure 10.1 depicts the process of bio gas plant.

The biogas contains chemical energy in the form of methane (CH4) which is the main part of biogas and is combustible.

10.2.2 Bio Gas Plant Components [3]

The main components of the bio gas plant as seen in Fig. 10.2 are:

1. **Mixing Tank**—Cow dung and water are mixed in equal proportion (1:1) to make a homogenous mixture called slurry. This mixing is done in the mixing tank. Since, biogas can be produced from various types of biomass, apart from cow dung we can use other wastes as dried leaves and agricultural crop waste.
2. **Digester Tank—**Digester Tank is where anaerobic digestion takes place and bio gas is formed. The tank is made of metal, reinforced concrete or plastic structure in which biochemical reactions occur with biomass fermentation. The digester tank can be underground, elevated or half-submerged. Fermentation creates ideal conditions for biomass fermentation and collection of biogas in the gasholder. For the fermentation process, the temperature is maintained by the heating system. The substrate in the digester is mixed with stirrers or agitators.
3. **In let Pipe**—Pipe connecting Mixing tank and digester, through which the slurry is introduced into the digester.

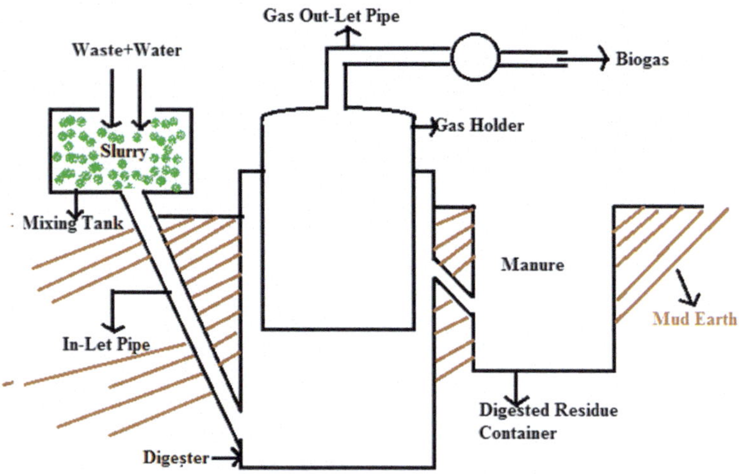

Fig. 10.2 Bio Gas Plant

4. **Out let Pipe for Bio Gas**—Pipe through which Bio gas is released and can be sent for further processing and application.
5. **Gas Holder**—Gas produced due to anaerobic fermentation is stored in gas holder. The gas holder is a drum in the case of *Khadi and Village Industries Commission* (KVIC)– biogas plant and is a dome in the case of fixed dome biogas plants. Gasholders can be low, medium or high pressure type. For efficient operation of the plant, the gas holder must be leak-proof and resistant to pressure and temperature.
6. **Digested Residue container**—Digested residue container is to un-load or extract the digested waste which is an organic fertilizer.
7. **Stirrer**—Stirrer, mixeragitator is placed in the digester for constant stirring and uniform chemical reaction. Stirrer are fixed verticallyor horizontally. The choice of type of Stirrer depends on the technology, there are mechanical, hydraulic or pneumatic. Stirring rate of operation effects the efficiency of the digester. In the present day we have various control techniques to monitor the speed of the stirrer. In the simple diagram of biogas plant in Fig. 10.2 stirrer has not been shown. However, Fig. 10.3 depicts a stirrer. The stirrer is a rod connected to the rotor of a motor. At the free end of the stirrer are blades as that of a fan.
8. **Gas Outlet Pipe**—The generated andstored gas is released through the gas outlet pipe present at the top of gas holder.

The digester capacity is designed based on the output energy required and the availability of the inputbio mass waste. All these components have to be designed based on the output. There are various models of the bio gas plants, and each model has all the above mentioned seven components and additionally several measurement and control systems attached. The classification and design consideration of the various types of bio gas plants is elaborated in the subsequent sections of this chapter.

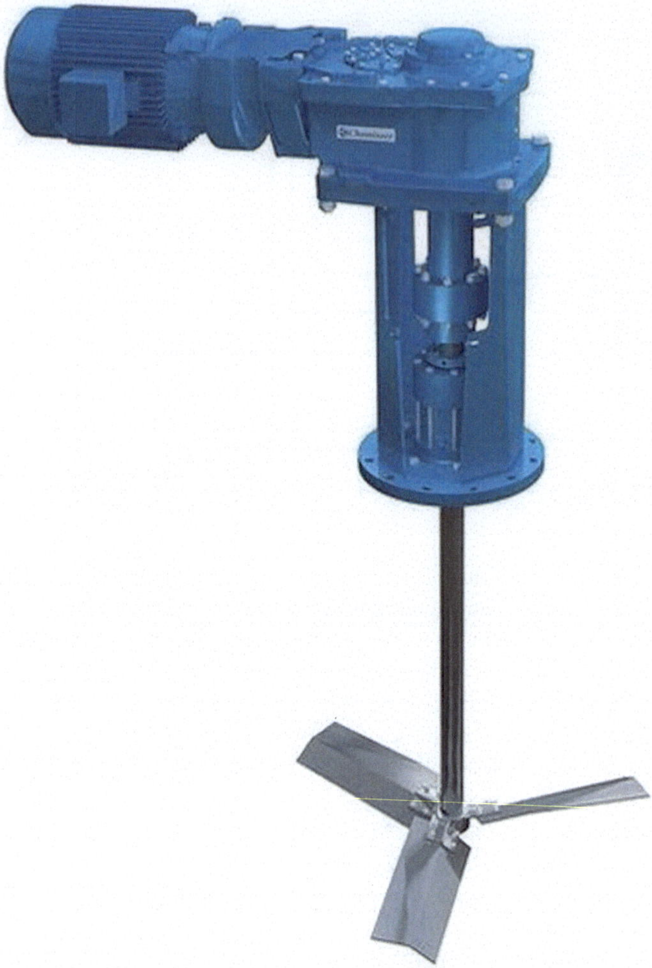

Fig. 10.3 Stirrer for biogas plants. (*Image courtesy*-**Sustain Energy Projects Private Limited, Chennai, Tamil Nadu**)

Apart from the above main components of a bio gas plant, the allied components in the plant for ease of operation are:

i. **Pre-*receiving tank is*** a tank where raw materials are treated before being fed into the digester to ensure their optimal fermentation.Its volume and design depend on the size, type and timing of the supply of raw materials that can be fed to the digester.
ii. Dome is a protective hood located above the gasholder and protects it from the effects of the atmosphere. The dome has high UV resistance, it is stretchable and elastic.

iii. ***Cogeneration module*** is a system used to burn biogas (with methane content CH_4 50–70%) at the appropriate temperature toremove the moisture and to generate heat and electricity.
iv. A gas burner is required for the combustion of low-quality biogas at the start of biogas plant operation and for the combustion of excess biogas.
v. Gas blowers are a type of air injection device located between compressors and fans in terms of pressure level. They are used to create transport flow paths for biogas.
vi. Biomix pump is special equipment for feeding the substrate used in biogas plants. The main purpose of the pumps is to supply solid material.The solid substrate is crushed and fed into the pump through a screw feeder.

10.3 Taxonomy of Bio Gas Plants

The various bio gas plants [4] are classified based on several criteria as in Table 10.1.

10.3.1 Batch Type Bio Gas Plant

Batch plants are filled at one time with a specific amount of biomass to full capacity. After retaining the biomass for a fixed retention time, the biogas generated is sent out through the outlet pipe. The digester is then completely emptied before filling for the next batch of operation.

Batch type biogas plants are suitable where daily supplies of raw waste materials are difficult to be obtained. Gas production is uneven as bacterial digestion starts slowly, peaks and then tapers off with growing consumption of volatile solids.

This problem can be overcome in two ways:

i. Having a number of digesters so that at least one is always in operation.
ii. Connecting batch loaded digester in series and scheduling their operation so that adequate biogas is available for daily use.

The important features of batch type biogas plants are:

(i) Gas production is uneven and interminent depending upon the clearing of the digester.
(ii) They have several digesters for continuous supply of gas.
(iii) Several digesters occupy more space.
(iv) The volume of digester is high, therefore, the initial cost is high.
(v) To start the digestion process, fermented slurry needs to be added.
(vi) They are suitable for long fibrous materials.

Table 10.1 Taxonomy of bio gas plants

Criteria	Type of plants	Principle of operation
Based on the rate of input feed	1. Batch	Batch plants are filled at one time with a specific amount of biomass. After retaining the biomass for a fixed retention time, the biogas generated is sent out through the outlet pipe. The digester is then completely emptied before filling for the next batch of operation
	2. Single stage biogas plant	As the name indicates, the process of conversion of biomass into biogas is carried out in a single chamber or digester
	3. Two stage biogas plant	In these plantsthere are two chambers for digestion of biomass **Stage 1**. Acid formation takes place and the diluted acids are only fed into the second chamber **Stage 2.** Methane formation occurs
	4. Continuous one stage	Continuous plants are fed continuously and emptied continuously, at a uniform rate as designed
Based on Digester Lid	1. Fixed Dome	In fixed-dome planta a closed, dome is fixed on top of the digester. The digester has an immovable, rigid gas-holder and a displacement pit
	2. Floating Dome	Floating-drum plants consist of an underground digester with a moving gas-holder. The gas-holder floats directly on the fermentation slurry or in a water jacket of its own
Additional Configurations	1. Plug-Flow Systems	Plug flow digesters have a constant volume, and produce biogas at variable pressure
	2. Anaerobic sequencing batch reactor (ASBR)	Anaerobic sequencing batch reactors (ASBR) as the name indicates have four cyclic steps: i. Feed, ii. Reaction, iii. Settling and iv. Discharge
	3. Tubular Reactor	Tubular reactors operation is a continuous flow with reagents flowing in and products being removed. They are the simplest of all reactor designs
	4. Anaerobic Baffled digesters or reactor (ABR)	The anaerobic baffled reactor (ABR) has a series of compartments. The number of compartments could be up to 8 in number in one reactor. It is an improved septic tank
	5. Upflow anaerobic sludge blanket reactor (UASB	The upflow anaerobic sludge blanket reactor (UASB) has a single tank, in which wastewater enters the reactor from the bottom, and flows upward

10.3.2 Continuous Type Bio Gas Plant

In continuous type biogas plant, feeding of the digester with biomass and the supply of the gas is continuous. Continuous biogas plants may be:

i. **Single Stage**—It has a single digester. These plants are economic, simple in construction and easy to operate. They generallyof small and medium size biogas plants.

   ii. **Double Stage or Multi stage**—There are two digesters, for Acidogenic and methanogenic stages. The first stage of acid production is carried out in a separate chamber and only diluted acids are fed into the second chamber where bio methanation takes place.

They are costlier, difficult in operation and maintenance as compared to single stage, but they produce more gas. These plants are suitable for larger biogas plant system.

The features of continuous type biogas plants are:

(i) Gas production is continuous.
(ii) Retention period is less
(iii) Less problems as compared to batch type.
(iv) Small digestion chambers are required

10.4 Fixed Dome Type Bio Gas Plants [4, 5]

A fixed-dome plant, as in Fig. 10.4 consists of a digester with adome-shaped cover which is fixed. When gas production commences, the slurry is displaced into the compensating tank. Gas pressure increases as the volume of gas stored increases. The plant is constructed underground, therefore saves space and free from any damage.

The costs of a fixed-dome biogas plant is lowand economical as:

- It is simple in construction.
- Has no moving parts.
- There are also no rusting parts and hence has long life of about20 years or more.

A models of fixed dome biogas plants are:

i. Deen Bandhu model biogas plants.
ii. Janata model biogas plants.

10.4.1 Deen Bandhu Biogas Plant

- Deen Bandhu model, Fig. 10.5 was developed in 1984, by Action for Food Production (AFPRO), a voluntary organization based in New Delhi.

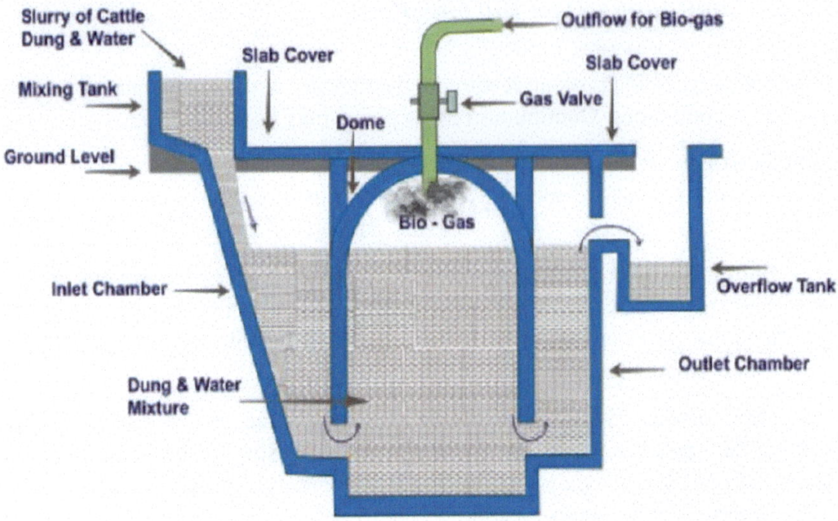

Fig. 10.4 Fixed dome plant

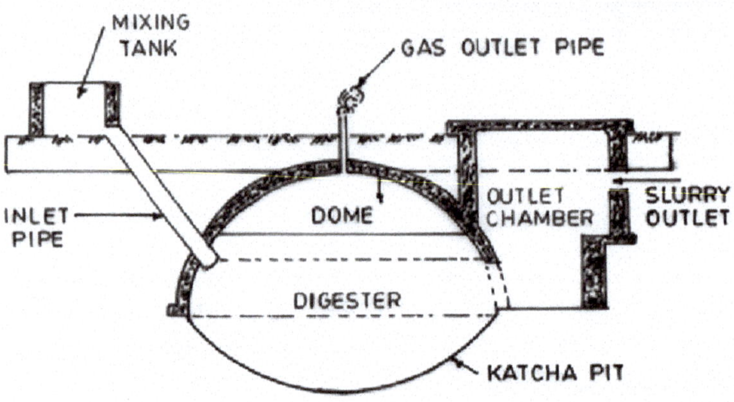

Fig. 10.5 Deenbandhu Bio Gas Plant

- Cost of the plant is very low, hence within the reach of even the poorer sections of the population.
- The cost reduction has been achieved by minimizing the surface area through joining the segments of two spheres of different diameters at their bases.
- The dome is made from pre-fabricated ferro cement or reinforced concrete and attached to the digester, which has a curved bottom.
- The slurry is fed from a mixing tank through an inlet pipe connected to the digester.

- After fermentation, the biogas that collects in the space under the dome is taken out for use through a pipe connected to the top of the dome. The sludge, which is a by-product, comes out through an opening in the side of the digester.

10.4.2 Janata Model Biogas Plants

This was first developed by the planning, Research and Actiondivision, Lucknow, in 1978. It is an improved version of the Chinese fixed domebio gas plant. The plant is shown in Fig. 10.6 with different parts. Digester is cylindrical in shape below ground level and is constructed with bricks and cements. At almost center of the digester, there are two rectangular openings facing each other and little above the ground level. These openings act as an inlet and outlet of the plant. The pipe fitted at the top of the dome is the gas outlet of the plant. The principle of gas production is same as that of KVIC model. The biogas is collected in the restricted space of the fixed dome, hence the pressure of gas is much higher, which is around 90 cm of water column.

This type of **biogas plant** is very economical is design. It works with the constant volume principle. This type of plant doesn't have any moving parts so it is free from wear and tear, and thereby easy maintenance. This model has longer life than KVIC model. Substrates other than cattle dung such as municipal waste and plant residues can also be used in Janata type plants.

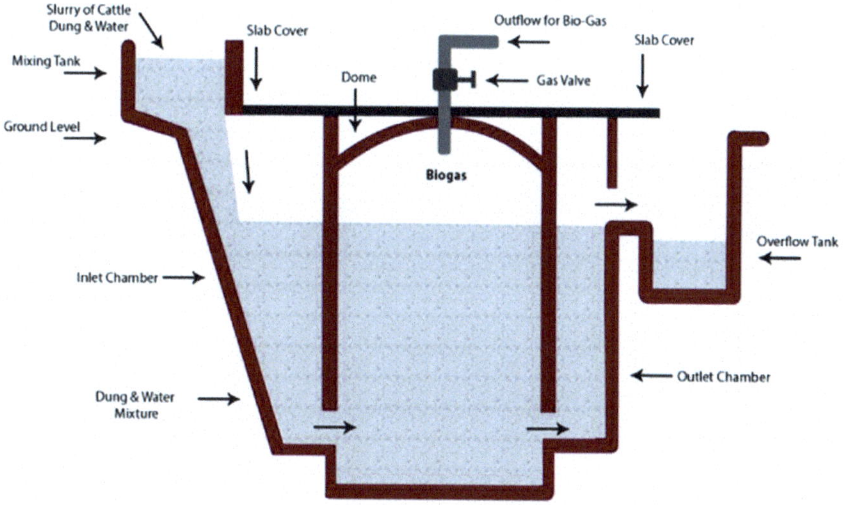

Fig. 10.6 Janata model biogas plants

10.5 Floating Dome Type Biogas Plants

Floating drum biogas plant is also known as movable drum type plant. The first floating drum biogas plant was developed in 1956 by Jashu Bhai J Patel of India. This is popularly called Gobar gas plant. Floating-drum plants consist of an underground digester which could be cylindrical or dome-shaped and a moving gas-holder. The gas collects in a steel gasholder which is inverted over the slurry and moves up and down depending upon accumulation and discharge of gas, hence it is known as floating drum.

Advantages of Floating Dome Plants:

(i) Constant gas pressure.
(ii) No problem of gas leakage
(iii) Higher gas production
(iv) Scum problem is less

Floating dome bio gas plants models are:

i. KVIC-type (KVIC- Khadi Village Industries Commission) biogas plant
ii. Pragati Design biogas plant
iii. Ganesh biogas plant

10.5.1 KVIC Type Biogas Plant

KVIC type bio gas plant consists of a digester and a floating drum for the collection of gas. Figure 10.7 depicts the Schematic diagram of a KVIC biogas plant.

Digester is 3.5–6.5 m in depth and 1.2–1.6 m in diameter. There is a partition wall in the center of the digester, which divides it vertically and submerges in the slurry

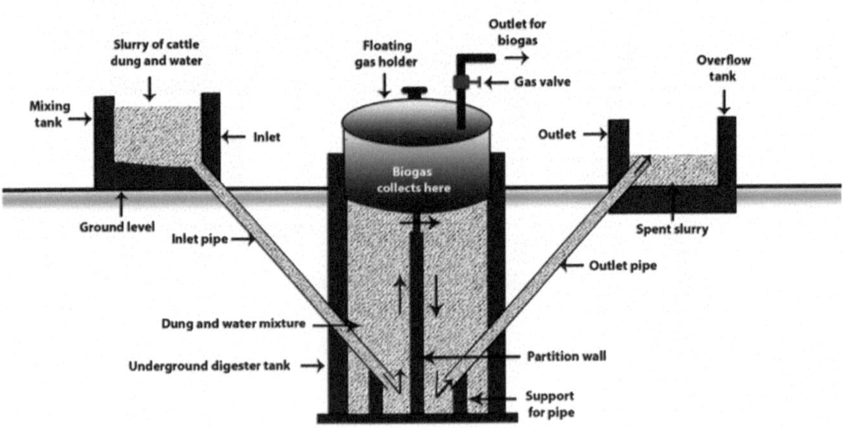

Fig. 10.7 KVIC biogas plant

when it is full. The digester is connected to the inlet and outlet by two pipes. Through the inlet, the dung is mixed with water in the ratio 4:5 and loaded into the digester. The fermented material will flow out through outlet pipe. The outlet is generally connected to a compost pit, where the digested slurry accumulated can be used as a fertilizer. The gas generation takes place slowly and in two stages. In the first stage, the complex, organic substances contained in the waste are acted upon by a certain kind of bacteria, called acid formers and broken up into small-chain simple acids. In the second stage, these acids are acted upon by another kind of bacteria, called methane formers and produce methane and carbon dioxide.

The gas pressure varies between 7 and 9 cm of water column. The unit cost of KVIC model with a capacity of 2 m^3/day costs approximately Rs. 14,000–00. The calorific value of bio gas ranges from 1600 to 2500 kJ/m3. It is a good fuel for cooking and lighting. When compared with diesel it is also be a very good fuel for compression, ignition engines and can save 70 to 80 percent of diesel.

10.5.2 Pragati Design Biogas Plant

Pragati Design Biogas Plant Developed by United Socio- Economic Development and Research Programme (UNDARP) Pune. Figure 10.8 depicts the Schematic diagram of a Pragati.

Design biogas plant.In Pragati Design biogas plant the depth of the pit is less than KVIC plant soits cost is 20% less and can be constructed in hilly and high water table locations.

10.6 Comparison of Deenbandhu and Kvic Bio Gas Plants [6]

See Table 10.2.

10.7 Simple Model of Bio Gas Digester for Domestic Cooking—Home Bio Gas

Figure 10.9 depicts the simple model of bio gas digester made of plastic for domestic cooking. The kitchen waste only can be used as input bio mass for the digester. The output gas can be connected to a cook stove.

Such models are also called home bio gas and can be made easily with the waste containers such as plastic drums available in any residence or organization.

The setting up conditions for a home bi gas are:

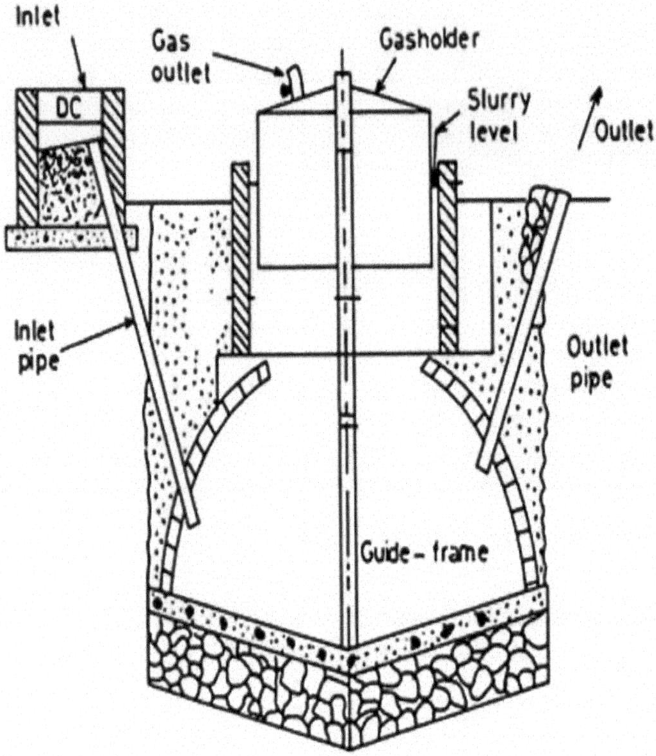

Fig. 10.8 Pragati Design biogas plant

Table 10.2 Comparison of Different BioGas Model w.r.t design, cost, shape

Deenbandhu bio gas plant	KVIC bio gas plants
Fixed dome type masonary structure	Floating dome type masonary structure
Under groundmasonary structure	Above ground masonary structure
Space above ground can be utilized for other beneficial purpose	Space above ground is used to fit movable drum
Initial investment is high	Initial investment is low
No maintenance cost	Incures some maintenance cost
Gas pressure is variable	Gas pressure is constant

i. HomeBiogas should be placed outdoors only, in a place where it can get maximum sunlight (not in the shade).
ii. On a stable floor, free from any obstructions.
iii. Water source should be close to where you intend to place the system.
iv. An average day/ night temperature of at least 20C/ 68F should be there.

Fig. 10.9 Simple model of bio gas digester for domestic cooking

There are many manufacturers and suppliers of home biogas plant at reasonable costs.

10.8 Modern Technology to Increase Efficiency of Bio Gas Plants

The efficiency of any system can be improved by making it a closed loop system. The closed loop system has various sensors that measure several parameters, The measured parameters enable determine the process state and control the input and dependent parameters of the system. In the context of modern technology, bio gas plants have developed with Internet-of-Things technology. Hence, the system process state can be monitored from remote locations also.

Effective biogas process optimizationrequire monitoring of the key parameters, methane, carbon dioxide, humidity, slurry consistency, temperature. In the past, monitoring the parameters effectivelyhad not been possible, but now with development of technology from various angles we canobtain more value from waste—improve efficiency, improve the profitability of biogas plants; help reduce waste; lower GHG emissions and recycle agricultural nutrients.

In this section we shall present the role of modern technology concepts in bio gas plants.

10.8.1 Sensors

Application

Sensors are used to measure the following operational parameters in bio gas plants:

i. Temperature
ii. pH
iii. Levels of feed going in—Input slurry and water ratio which is Organic Loading Rate (OLR)
iv. Hydraulic Retention Time (HRT)
v. Dry Matter
vi. Organic Dry Matter (ODM)
vii. Pressure
viii. Volatile Fatty Acids
ix. Bio gas output rate
x. Bio gas output composition of Methane (CH_4), Carbon Dioxide (CO_2), Hydrogen Sulphide (H_2S), Hydrogen (H_2)
xi. Viscosity
xii. Carbon/Nitrogen (C/N) Ratio
xiii. Oxidation Reduction Potential (ORP)
xiv. Agitation—Stirring of the digester slurry.
xv. Electrical Conductivity measurements on real time basis.
xvi. Fourier Transform Infra-Red (FT-IR) spectrometer can be installed in the digesters to detect and determine

    (a) Chemical Oxygen Demand (COD),
    (b) Total Organic Carbon (TOC),
    (c) Volatile Fatty Acids (VFA), and
    (d) Partial and Total Alkalinity (PA and TA).

Ward et al. (2011) [7] conducted a study in a pilot scale biogas plant to examine the feasibility of measurements of Micro Gas Chromatography (μ-GC), Membrane Inlet Mass Spectroscopy (MIMS) in gas phase and with Near Infrared Spectroscopy (NIRS), pH in the liquid phase. The results showed that μ-GC is suitable to determine Hydrogen(H_2), Methane(CH_4), Hydrogen Sulphide(H_2S), Nitrogen(N_2) and Oxygen(O_2) and MIMS can be used for determining Methane(CH_4), Carbon Dioxide(CO_2), Hydrogen Sulphide(H_2S).

Online automated systems would enable sufficient monitoring at the biogas plants, especially in the rural areas [8]. The currently available technologies do not support the monitoring of all operational parametersof the biogas plant; measure and process the parameters to take immediate control actions. As on-line monitoring is not yet completely developed, samples have to be collected from the biogas plant and analyzed in individual facilities (off-line monitoring). The samples are tested in the laboratory for various relevant tests. However, the development of Internet-of-Things(IoT) and Microcontrollers has provided means and methods for the monitoring and control of biogas plant operation to a certain extent.

Types of Sensors

Sensors can be classified based on their principle of Operation and physical distance from the process monitored component as waste materials [9]

The types of sensors in the two categories are presented in Table 10.3.

Table 10.3 Categories of Sensors

Sensors based on their principle of Operation	Sensors based on the physical distance from the process monitored component
1. Photonic Sensors	Contact Sensors as: • Resistivity sensors, • Time-domain reflectrometry (TDR) sensorsand • Neutron particle probe sensors, can be embedded into the waste samples. Thus giving good results
2. Radiowave Sensors	Proximity Sensors as: • Open-ended coaxial line probe (OECP) radiowave sensors, • X-ray sensors, • Capacitance sensors, and inductive sensors, must operate in near-contact with the waste samples
3. Electric and Magnetic Sensors as Capacitive sensors, Resistive sensors, Inductive sensors, Neutron sensor probe	Remote Sensorswith: • Ultraviolet (UV) to long-wavelength infrared (LWIR) spectroscopy, • UV to short-wavelength infrared (SWIR) imaging, • Radiowave permittivity measurements, can be operated with variable distances to the waste samples, providing flexibility in monitoring and control
4. High-Energy Particle Sensors	

10.8.2 Digital Signal Processors

Technologies such as Digital Signal Processors can be custom built for a particular bio gas plant design. On-line monitoring provides real-time data on the biogas plant's operation. A study about the on-line monitoring system was done in 2013 in Germany, and it was observed that the majority of the biogas plants are equipped with on-line systems to monitor the biogas plant operation and output electricity generation. To improve monitoring systems of the biogas plants, the best technologies are—near infrared spectroscopy (NIR) and mid-infrared spectroscopy (MIR).Therefore, In order to obtain operation flexibility of biogas plants as adjusting operational parameters, upgradation of biogas plants by integrating the new technologies is a good policy.

10.8.3 Artificial Intelligence (AI)

Several studies regarding implementation of AI and the modern day data analytics in the area of living (biological) systems, have been conducted and considerable results have been obtained. In dynamic condition systems such as biogas plants, where there are not only internal parameters but also external parameters affecting the process, real time monitoring and control is of great importance to balance the chemical reaction and obtain maximum output. Wahmkow has studied the usage of neural networks and fuzzy logic in his work. As an optimum solution to build a driver for biogas production, a combination of Neural Network, Fuzzy Logic and optimization was suggested. In the end it was claimed that with the usage of automated controller techniques, technical and ecological goals could be achieved. Bhuiyan et al. (2019 have studied the feasibility of having a sustainable and affordable way to produce renewable energy at any scale for all end level users in the world. In this model it was aimed to find a new method to produce clean energy with the help of modern technology like IOT and AI. Moreover, Olabi [10] aimed at maximizing the methane production from wastepaper via the integration of AI to the monitoring.

Fuzzy logic-based modeling and modern optimization were proposed by several researchers and found that implementation of fuzzy model, increased methane yield.

Oznur Yildirim [11] et.al in their study "Prediction of biogas production of industrial scale anaerobic digestion plant by machine learning algorithms" have implemented five different machine learning (ML) algorithms—RF, **ANN**, KNN, SVR, and XGBoost to describe and predict the correlation between operational parameters and biogas production quantities collected from a real-scale anaerobic digestion plant. Figure 10.10 depicts the various operational parameters monitored and the Al Algorithms implemented in the process control of the biogas plant. They observed that KNN algorithm had the lowest accuracy in predicting total biogas production over time, while the RF model had the highest prediction accuracy of all prediction models. The RF method produced the best prediction, with an R^2 of 0.9242, and

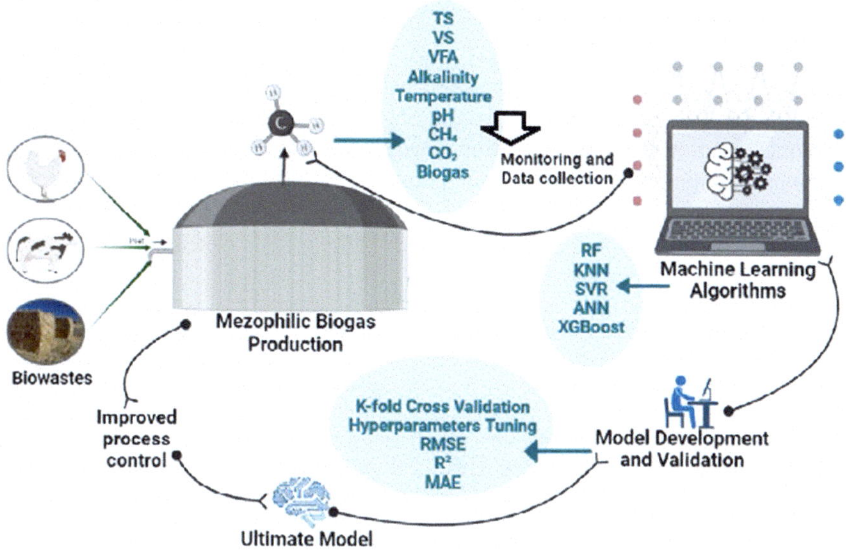

Fig. 10.10 Implementation of Machine Learning in Biogas Plant

it was followed by XGBoost, ANN, SVR, and KNN (with R^2 values of 0.8960, 0.8703, 0.8655, 0.8326, respectively). Real-time process control has thus enabled maintain process stability and improve biogas production with the integration of ML applications into AD facilities.

Artificial Neural Networks are thus becoming an important tool in the field of biogas production.

10.8.4 Industry 4.0 [12]

Industry 4.0is the fourth industrial revolution, which is a combination of several emerging technologies working together to provide more efficient and faster process control. Industry 4.0 and Industrial Internet of Things (IIoT) technologies provide data and software solutions based on digitalization in many areas, especially industrial automation and production systems. The problems of AD systems can be overcome by integrating next-generation Artificial Intelligence techniques and industry 4.0 systems into biogas plants. This integration of new software and hardware technology would- increase the efficiency of the system, provide on-site control, improve process performance in AD facilities and thereby also make system operation and maintenance easy.

10.8.5 Chemical Techniques

Chemical techniques change the chemical constitution and composition of waste. The primary techniques are thermal and non-thermal techniques. Now advanced modern techniques, such as hydrate separation, cryogenic separation, biological methods, membrane enrichment, in-situ upgrading, multistage and high-pressurized anaerobic digestion have been developed.

10.9 Digester Operational Parameters [13]

The operational parameters if monitored would lead to efficient operation of the biogas plant. The monitoring of the biogas plant's operational parameters can be done by on-line, and off-line modes. The development in various areas of technology as microprocessors, PLC and various types of sensors has enabled minute-to-minute monitoring and control of the process. There are several parameters that need to be monitored at the biogas plant on a real-time basis.

The operational parameters of the biogas digester for its effective and efficient functioning are:

1. **C/N ratio:** C/N ratio is the ratio of carbon to nitrogen present in the feed material.For proper anaerobic digestion to take place the C/N ratio ranges from 20 to 30:1.
2. **Temperature:** Rate of reaction inside the digester is dependent on temperature inside and outside the digester. The best temperature range is 350–380 °C, and as the ambient temperature increases the rate of reaction increases, thus enhancing the biogas production.
3. **Retention time:** It is the time that period for which the feedstock remains in the digester for complete digestion. It is computed as the **volume of digester divided by the feedstock added per day**. Retention Time unit is days and calculated as:

$$R_T = V_R / V \tag{10.1}$$

where $R_T =$ retention time $\left[\text{days}\right]$
$V_R =$ digester volume $[m^3]$.
$V =$ volume of substrate fed per time unit $[m^3/d]$.

Example: 32 kg of cow dung + 32 l water = 64 l fermentation slurry.

The digester volume therefore is:
Digester volume (l) = Daily feed (l/day) × Retention time (days)
Assuming the Retention time to be 50 days, then the digester volume can be calculated by.
Digester volume = 64 (l/day) × 50 (days) = 3200 l or 3.2 m^3

4. **Loading rate**: Loading rate is the amount of raw material that is fed to the digester per day per unit volume

   Loading Rate or Organic Loading Rate is calculated as:

$$BR = m \times c/VR \tag{10.2}$$

where: BR is the organic load in kg/d*m^3
   m is the mass of substrate fed per unit time i.ekg/day
   c is the %concentration of organic matter
   VR is the digester volume in m^3

5. **pH or hydrogen ion concentration**: For constant generation of gas in the digester, it is necessary to maintain pH of the slurry between 6.5 and 7.5.The pH of the slurry changes as the digestion progresses.
6. **Total solid content**: The raw cow dung contains 80–82% of moisture and 20–18%% of total solids. The cow dung is mixed usually in the proportion of 1:1 in order to bring the total solid content to 8–10%.
7. **Diameter to depth ratio**: The diameter to depth ratio for maximum gas production has been found as per studies to be in the range of 0.66–1.00.
8. **Degree of mixing**: Degree of mixing is the rate of stirring to make the slurry uniform. Mixing improves fermentation, but violent slurry agitation retards the digestion.

10.10 Calculating the Daily Gas Production (G)

Daily biogas production can be estimated using the following equation (Fulford, 2015):

$$G = C \times Vd \times S \times \left[\frac{K}{1 + KR}\right] \tag{10.3}$$

where:

- G is the biogas production (in m^3/day)
- C is the biogas potential. It is the maximum amount of gas that can be produced from 1 kg of volatile solids in a feedstock (in m^3/kg)
- Vd is the digester volume (in m^3)
- S is the initial concentration of volatile solids in the slurry (in kg/m^3)
- R is the feedstock retention time (in days)
- k is a constant. It indicates the rate of gas production at a given temperature

Note: The above equation can be used for any one feedstock.

To simplify this equation, International Renewable Energy Agency (IRENA) has calculated the gas production across a wide range of temperatures and retention times. Table 10.4 gives the Yield factors for biogas production, by temperature and feedstock retention time,

so that biogas production can be calculated as follows:

$$G = \frac{(Y \times Vd \times S)}{1000} \tag{10.4}$$

where Y is a yield factor based on temperature and the feedstock retention time.

$$\text{For example, If yield factor (Y)} = 4.00,$$

$$\text{digester volume (Vd)} = 5.0\ \text{m}^3 \text{ and}$$

$$\text{initial concentration of solids (S)} = 80\ \text{kg/m}^3,$$

Table 10.4 Yield factors for biogas production, by temperature and feedstock retention time

Feedstock retention time (in days)	Temperature (°C)					
	16–18	19–21	22–24	25–27	28–30	31–33
6–10	5.41	7.98	10.83	13.59	15.91	18.33
11–15	4.73	6.79	8.99	11.09	12.88	14.74
16–20	4.21	5.90	7.68	9.37	10.82	12.32
21–25	3.79	5.22	6.70	8.11	9.33	10.59
26–30	3.44	4.69	5.95	7.15	8.20	9.28
31–35	3.16	4.25	5.35	6.39	7.32	8.26
36–40	2.91	3.88	4.86	5.78	6.60	7.44
41–45	2.71	3.58	4.45	5.27	6.02	6.77
46–50	2.53	3.32	4.10	4.85	5.53	6.21
51–55	2.37	3.09	3.81	4.49	5.11	5.74
56–60	2.23	2.89	3.55	4.18	4.75	5.33
61–65	2.10	2.72	3.33	3.91	4.44	4.98
66–70	1.99	2.57	3.13	3.67	4.17	4.67
71–75	1.89	2.43	2.95	3.46	3.93	4.40
76–80	1.80	2.30	2.80	3.27	3.71	4.15
81–85	1.72	2.19	2.66	3.10	3.52	3.94
86–90	1.65	2.09	2.53	2.95	3.34	3.74
91–95	1.58	2.00	2.41	2.81	3.19	3.56
96–100	1.52	1.92	2.31	2.69	3.04	3.40

Courtesy: IRENA-2016.

$$\text{daily gas production } G = \frac{(4 \times 5 \times 80)}{1000} = 1.6\,\text{m}^3/\text{day}$$

10.11 Techno-commercial Analyses of Biogas Plants in India

Numerous empirical studies and research in India and abroad carried out by various agencies have yielded the following results in general.

- One ton (1000 kg) of biodegradable waste yields nearly fifty cubic metres (50m^3). of biogas per day. The yield depends on various factors like temperature in the digester, characteristics of the slurry material and several operating parameters.
- One cubic meter (1m^3) of biogas is equivalent to four hundred grams (0.4 kg) of LPG in accordance with the ratio of calorific value of biogas to that of LPG.
- One ton of biodegradable waste generates nearly 100 kg of manure as by product.
- The biogas plant of capacity 1ton per day in India costs Rs.1.5 Million

Considering the above parameters, the economics of biogas plant to manage 75,000 Tonnes of waste in India is summarized in Table 10.5.

Table 10.5 Commercial Analysis of Biogas Plant as a Solution to Manage Biodegradable Waste in India

Parameters	Daily	Yearly
Total quantity of waste (in TPD) that be managed through biogas plants	75,000.00	27,375,000.00
Biogas generated (Cubic metre)	3,750,000.00	1,368,750,000.00
Biogas equivalent of LPG (kg)	1,500,000.00	547,500,000.00
Savings through LPG per day (Rs. in Million)*(a)	134.34	49,034.10
Manure generation (in TPD)*(b)	7500.00	2,737,500.00
Savings through manure (Rs. in Million)	22.50	8,212.50
Total savings through biogas and manure (Rs. in Million)	156.84	57,246.60
Investment (Rs. in Million)	112,500.00	

* The cost of LPG is considered as Rs.89.56/kg as per the prevailing market rates

A biogas bus, Sweden

The Biogas Train "Amanda" Sweden

Indian Oil Retail Outlet at Talegoen,Pune,India selling Compressed Bio Gas.

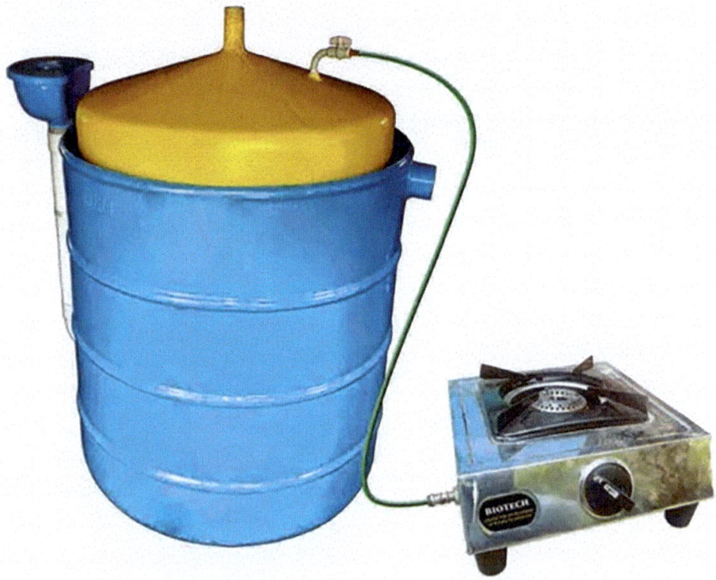

Domestic Biogas Plant for Cooking

References

1. https://www.nationalgrid.com/stories/energy-explained/what-is-biogas
2. https://www.rumahenergi.org/en/2019/11/27/electricity-from-biogas.html
3. https://www.powerup.at/knowledge/biogas-plant/components/
4. https://energypedia.info/wiki/Types_of_Biogas_Digesters_and_Plants
5. https://www.indiastudychannel.com/resources/117400-various-types-biogas-plants.aspx#:~:text=BATCH%20TYPE%20BIOGAS%20PLANT,is%20emptied%20and%20filled%20again
6. Comparison among different models of biogas plants, June 2015, https://doi.org/10.13140/RG.2.1.1624.6882, Project: Biomass Conversion Research Centre, COMSATS University Islamabad
7. [https://www.mdpi.com/2227-9717/9/1/85, "Integration of Artificial Intelligence into Biogas Plant Operation" by Samet Cinar, Senem Onen Cinar, Nils Wieczorek, Ihsanullah Sohoo And Kerstin Kuchta "Sustainable Resource and Waste Management" https://doi.org/10.3390/pr9010085]
8. https://d-nb.info/1216573522/34, Operational Parameters of Biogas Plants: A Review and Evaluation Study Abdullah Nsair, Senem Onen Cinar , Ayah Alassali , Hani Abu Qdais and Kerstin Kuchta,Published: 22 July 2020, Energies 2020, 13, 3761; https://doi.org/10.3390/en13153761
9. Sensor Technology Options for Municipal Solid Waste Characterization for Optimal Operation of Waste-to-Energy Plants Harald Ian D. I. Muri , and Dag Roar Hjelme, Energies 2022, 15, 1105. https://doi.org/10.3390/en15031105 https://www.mdpi.com/journal/energies
10. Olabi A, Nassef AM, Rodriguez C, Abdelkareem MA, Rezk H (2020) Application of artificial intelligence to maximize methane production from waste paper. Int J Energy Res 44:9598–9608
11. Oznur Yildirim, Bestami Ozkaya, Prediction of biogas production of industrial scale anaerobic digestion plant by machine learning algorithms, Chemosphere, Volume 335, 2023, 138976,

ISSN 0045–6535, https://doi.org/10.1016/j.chemosphere.2023.138976. (https://www.sciencedirect.com/science/article/pii/S0045653523012432)

12. "Development of an Industry 4.0-Enabled Biogas Plant for Sustainable Development",B.Rajesh Reddy et.al., Book on Advances in Industrial and Production Engineering
13. "Harnessing Solid Waste To Energy: A Techno-Commercially Feasible Solution for Effective Solid Waste Management in India", K. N. Narasimha et.al, International Journal of General Engineering and Technology (IJGET) ISSN(P): 2278–9928; ISSN(E): 2278–9936 Vol. 10, Issue 1, Jan–Jun 2021; 75–84© IASET

Chapter 11
Energy from Landfill Gas

11.1 Introduction

Landfill is an area where solid waste is dumped for further handling [1]. The further handling involves segregation of various type of wastes, compressing the wastes and dispatching. The waste after segregation is compressed for ease of handling and dispatched for further processing which could be re-cycling or combustion. Re-cycling process generated a new product, such as paper recycled can be converted to useful products Fig. 11.1 such as wooden boxes, greeting cards and carry bags. Glass bottle re-cycled can be converted to decorative articles. Waste can also be used to generate various forms of energy as thermal and electrical, for which it is transported to Waste-To-Energy Plants.

However, since many tons of waste is dumped in the landfill it would take considerable time to dispose. During this long duration wastes such as food products due to natural bacterial breakdown gases are generated and also leachate seeps into the soil. Figure 11.2 depicts the formation of leachate and gases from solid waste.

11.2 Landfill

A landfill site, Fig. 11.3, is also known as a tip area, dumping yard, or dumping ground, a site for the disposal of waste materials. Landfill sites [2] can be used for various purposes of waste management, such as temporary storage, consolidation and transfer point, segregation of various waste material, waste size reduction by compression or crushing or recycling. Leachate formed at the bottom of the waste on the soil should be properly handled, and landfills must be treated to avoid severe shaking or soil liquefaction of the ground during an earthquake.

N. Manohar et al., *Machine Learning Applications in Renewable Energy*, Green Energy and Technology, https://doi.org/10.1007/978-981-97-9939-8_11

Fig. 11.1 Re-cycling waste paper to useful products as greeting cards, paper bags and boxes

Fig. 11.2 Formation of leachate and gases from solid waste

11.3 Landfill Gas (LFG)

Landfill gas is generated by the natural bacterial breakdown of organic waste at municipal solid waste (MSW) landfills. It contains approximately 50% methane, with the remainder being mostly Carbon Dioxide (CO_2), with smaller amounts of water vapor, nitrogen, oxygen, and trace amounts of other contaminants. The gas is wet and also considered dirty.

When MSW is deposited in a landfill, it undergoes an aerobic decomposition stage and little methane is generated. Then, typically within less than 1 year, anaerobic conditions are established and methane-producing bacteria begin to decompose the waste and generate methane. Methane is a potent greenhouse gas at least 28 times more effective than CO_2 at trapping heat in the atmosphere.

11.3.1 How is Landfill Gas Produced? [3]

Landfill gas is produced by the following three processes:-

Fig. 11.3 Landfill site

i. Bacterial decomposition
ii. Volatilization, and
iii. Chemical reactions

i. **Bacterial decomposition.** Landfill gas is produced primarily by the bacterial decomposition of when organic waste broken down by bacteria naturally present in the waste and in the soil. Organic wastes include food, garden waste, domestic wastes, textiles, wood and paper products. Bacteria decompose organic waste in four phases. At each phase the composition of the gas changes. Figure 11.4 shows gas production at each of the four stages. Landfill conditions effect the generation of landfill gas, as duration of the waste on the landfill and temperature conditions.

ii. **Volatilization:** Landfill gases can be created when certain wastes, particularly organic compounds, change from a liquid or a solid into a vapor. This process is known as volatilization. Non Methane Organic Compounds (NMOC) in landfill gas may be the result of volatilization of certain chemicals disposed of in the landfill.

iii. Chemical reaction:. Landfill gas and NMOCs, can be created by the reactions of certain chemicals present in waste, for example if chlorine bleach and ammonia come in contact with each other in the landfill waste, a harmful gas is produced.

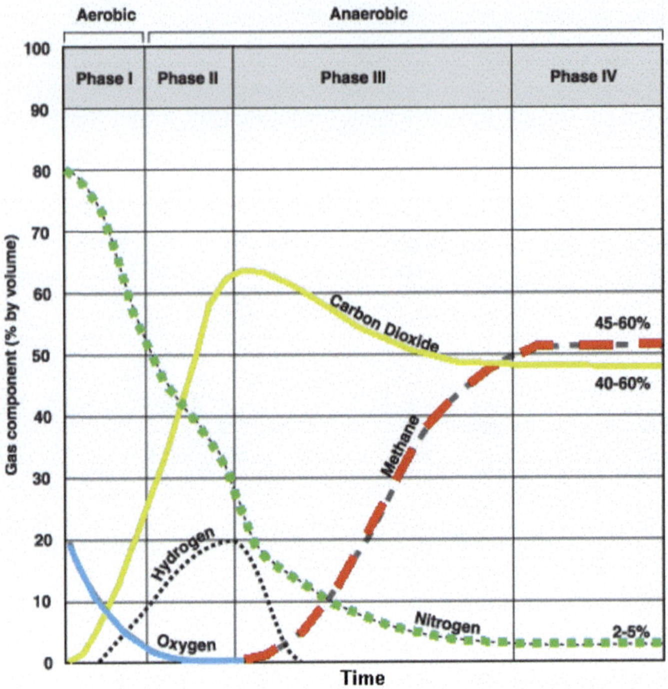

Fig. 11.4 Chemical changes in LFG over time

11.4 Collecting and Treating Landfill Gas

Instead of letting the gas escape into the air, LFG can be captured, converted, and used as a renewable energy resource. Use of LFG helps to protect the environment by reducing odors and other hazards associated with LFG emissions. In addition, LFG energy projects would also help the society by generating revenue and provide jobs for various category skill personnel.

Figure 11.5 illustrates the collection and processing of LFG to produce methane for multiple uses. LFG is first collected through vertical and horizontal pipes buried in an MSW landfill. The LFG is then processed and treated for use. Figure 11.5 also shows the potential uses of LFG for electricity generation, industrial and institutional uses, and as fuel for vehicles.

The four stages of LFG process and treatment are depicted in Fig. 11.6 and the real-time application and processing methodology in Fig. 11.7:

1. **Collection**: LFG is extracted from landfills using a series of wells and a blower/flare or vacuum systems. The collected gas is transported to a central location where it is processed and treated depending upon the ultimate use for the gas, to the point of consumption

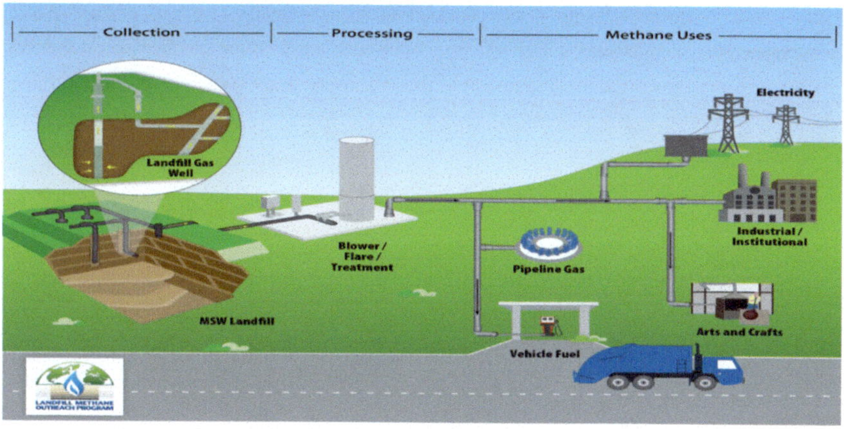

Fig. 11.5 Collection and processing of LFG to produce useful energy

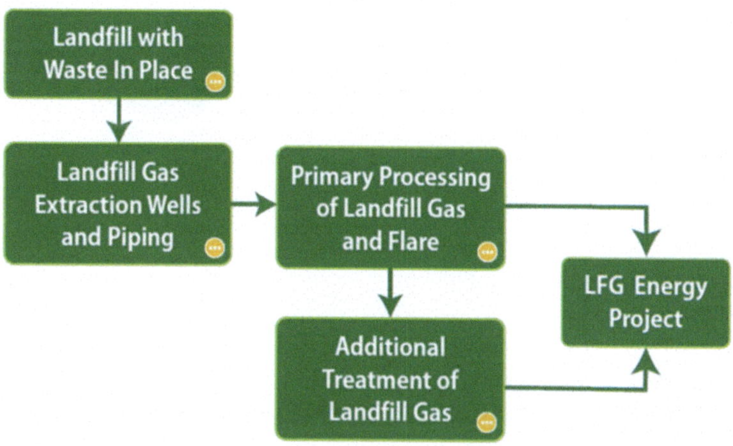

Fig. 11.6 Flow chart of LFG treatment

2. **Primary Treatment**: Removing moisture as the gas passes through a knockout pot, filter, and blower.
3. **Secondary Treatment**: In secondary treatment additional moisture is removed to the extent necessary by coolers. Then removal of siloxane/sulfur is done and compressed (as needed). After the impurities are removed in the Secondary Treatment stage, the LFG is transported for generation of electricity or as a medium-Btu fuel for boilers.

4. **Advanced Treatment:** Removing additional impurities as CO2, N2, O2, and VOCs and compress the LFG into a high-Btu gas that can be used for vehicle

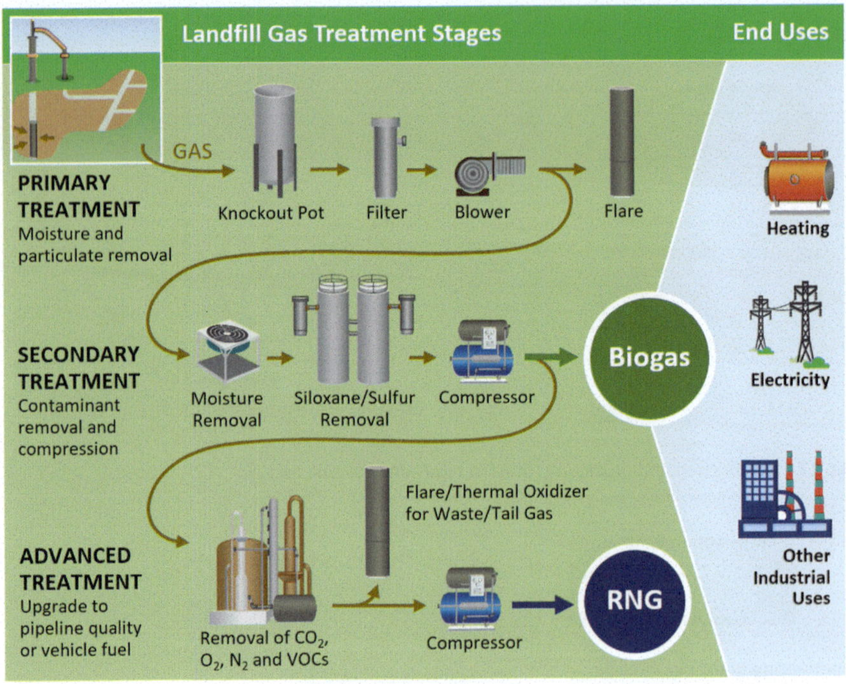

Fig. 11.7 LFG treatment and applications

fuel or injected into a gas pipeline. The residue tail gas is sent to a thermal oxidizer.

11.5 Conditions Affecting Landfill Gas Production

The amount of landfill gas produced at a specific site depends on several aspects related to the waste and environment as:

- The characteristics of the waste such as composition and age of the refuse and
- Number of environmental factors as the presence of oxygen in the landfill, moisture content, and temperature
- **Waste composition.** The more the organic matter present in the waste, the more landfill gas with components as carbon dioxide, methane, nitrogen, and hydrogen sulfide is produced. The more chemicals disposed of in the landfill, the more likely NMOCs and other gases will be produced through bacteria decomposition, volatilization or some chemical reactions.
- **Age of refuse**. Generally, more recently buried waste of less than 10 years produces more landfill gas than does older waste buried more than 10 years back. Peak gas production usually occurs from waste of 5–7 years old.

- **Presence of oxygen in the landfill.** Methane gas will be produced only when oxygen is no longer present in the landfill.
- **Moisture content.** The presence of moisture in a landfill increases gas production as it encourages bacterial decomposition. Moisture may also promote chemical reactions that produce various gases.
- **Temperature.** As the landfill's temperature rises, bacterial activity increases, resulting in increased gas production.

11.6 Landfill Gas to Energy [4]

The different types of LFG energy projects can be grouped into three broad categories:-

1. Electricity Generation
2. Direct Use of Medium-Btu Gas, and
3. Renewable Natural Gas.

11.6.1 Electricity Generation from Landfill Gas

Electricity can be produced on-site, using technologies like reciprocating internal combustion engines, turbines, micro turbines, and fuel cells. The excess energy generated can be sold to the grid. The reciprocating engine is widely utilized in LFG electricity applications due to its affordable cost, excellent efficiency, and ability to match the gas output of various landfills in terms of size. Gas turbines are commonly employed in bigger LFG energy projects, while micro turbines are generally utilized for smaller LFG projects and in specialized applications. Cogeneration projects, produce heat or thermal energy and electricity at a single plant utilizing LFG. Hence, cogeneration system is also known as combined heat and power (CHP) project. Many cogeneration projects have been implemented in industrial, commercial, and institutional operations, employing both engines and turbines. An appealing project type due to the potential to capture both thermal energy and generate electricity, resulting in increased efficiency.

Disposing of waste in landfills, the gas produced can either be burned off in a flare or used for energy in landfill-to-energy projects.

While it may be advantageous to measure the flow from every well, the expense associated with it makes it impractical. However, the same type of information can be obtained by using a portable device called a thermal mass flow meter, as shown in Fig. 11.8 periodically, the flow from the wellhead is sampled over time. This process is carried out repeatedly.

Accurate measurement of gas flow is crucial at every step of the process, from the extraction of gas from the landfill to its destruction or use in cogeneration. This is necessary for optimizing the process and for reporting greenhouse gas emissions

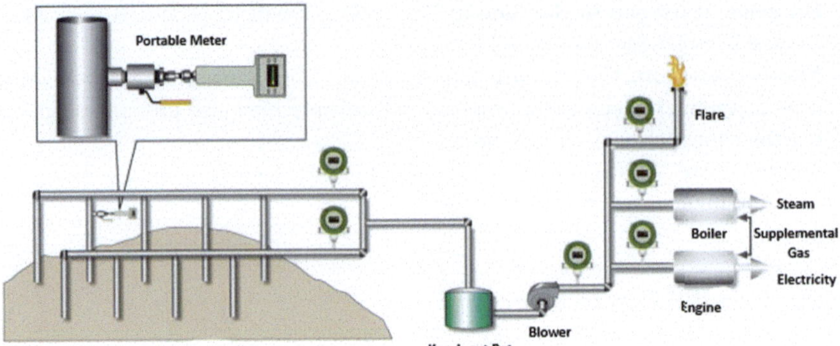

Fig. 11.8 Mass portable flow meter measuring flow from each well

to environmental agencies. Figure 11.8 shows a portable flow meter measuring the flow of gas from each well.

Certain local government landfills are unable to carry out landfill-to-energy initiatives. In such situations, the gas emitted from the landfill is typically gathered and transported to a privately owned co-generation facility, which is usually situated within the property. The purpose of this facility is to convert the gas into electricity. It is discharged or released during times when the co-gen facility doesn't require the gas. In both situations, it is necessary to use gas flow meters to calculate the flow of landfill gas towards the co-generation system or the flare. The purpose is to accurately measure the amount of landfill gas being produced.

Performance from any flow meter needs to be high due to operate effectively under difficult conditions such as:

- Variotions in gas and ambient temperatures
- Fluctuating gas flow rates
- Potentially hazardous and and explosive locations

11.6.2 Direct Use of Medium-Btu Gas

Directly using LFG can offset the use of another fuel (for example, natural gas, coal or fuel oil). LFG can be used directly in a boiler, dryer, kiln, greenhouse or other thermal application. In these projects, the gas is piped directly to a nearby customer for use in combustion equipment as a replacement or supplementary fuel. Only limited condensate removal and filtration treatment are required, although some modifications of existing combustion equipment might be necessary.

LFG can also be used directly to evaporate leachate. Leachate evaporation using LFG is a good option for landfills where leachate disposal at a water resource recovery facility is unavailable or expensive. LFG is used to evaporate leachate to a more concentrated and more easily discarded effluent volume.

Innovative direct uses of medium-Btu gas include firing pottery and glass-blowing kilns; powering and heating greenhouses; and evaporating waste paint. Current industries using LFG include auto manufacturing, chemical production, food and beverage processing, pharmaceuticals, cement and brick manufacturing, wastewater treatment, consumer electronics and products, paper and steel production, and prisons and hospitals.

11.6.3 Renewable Natural Gas

LFG can be upgraded to renewable natural gas (RNG), a high-Btu gas, through treatment processes by increasing its methane content and, conversely, reducing its CO_2, nitrogen and oxygen contents. RNG can be used in place of fossil natural gas, as pipeline-quality gas, compressed natural gas (CNG) or liquefied natural gas (LNG). About 16% of currently operating LFG energy projects are creating RNG.

Options for use of RNG include thermal applications, to generate electricity or as fuel for vehicles. The RNG can be used locally at the site where the gas is produced or can be injected into natural gas transmission or distribution pipelines for delivery to another location.

11.7 Advantages of Energy from Landfill Gas

- Reduction of greenhouse gas emissions
- Efficient use of waste
- Reduction in air pollution
- Improve air quality
- Cost Effective
- Generate energy at the point of waste generation and point of consumption
- Reduction in landfill odour
- Reduction in soil pollution due to reduction in leachate formation

11.8 Measurement of Landfill Gas [5]

Gas production over time may be estimated from the EPA Land GEM model based on Eq. 11.1

$$QT = \sum_{i=1}^{n} 2kLoMie^{-kti} \tag{11.1}$$

where:

QT = Rate of total gas emission from a landfill in volume/time.
n = Waste placement Total time periods.
k = Constant for landfill gas emission.
Lo = Generation of methane in volume/mass of waste.
ti = ith time period i.e. time for which waste is placed.
Mi = Mass of wet waste placed at time i.

Example 11.1 A landfill cell is open for five years, receiving 95,000 tons of waste per year. Find the peak gas production for the first year if the landfill gas emission constant is 0.03 per year, and the methane generation potential is 145 m^3/ton.

Solution

Given: Mi =95000 Tons/year
$k = 0.03$, $Lo = 145$, $t_i = 1$
For the first year,
$QT = 2\ (0.03)\ (145)\ (95{,}000)\ (e - 0.03(1)) = 802{,}073\ m^3$
Peak gas production for the first year = 802,073 m^3

References

1. https://sagemetering.com/applications/technical-notes/landfill-gas-monitoring-recovery-and-flaring/
2. https://www.epa.gov/lmop/basic-information-about-landfill-gas
3. Cheremisinoff NP (2003) Handbook of solid waste management and waste minimization technologies. Butterworth-Heinemann
4. https://inoplex.com.au/information/how-is-landfill-gas-turned-into-electricity/#:~:text=The%20captured%20landfill%20gas%20is,create%20electricity%20for%20various%20purposes
5. Abdel-Magid I (2012) Problem solving in solid waste engineering. https://doi.org/10.13140/RG.2.1.1251.6327

The manufacturer's authorised representative in the EU is Springer Nature Customer Service Centre GmbH, Europaplatz 3, 69115 Heidelberg, Germany. If you have any concerns regarding our products, please contact ProductSafety@springernature.com

Printed and bound by CPI Group (UK) Ltd, Croydon, CR0 4YY

30/09/2026

02225302-0005